STRUCTURE OF ATOM

STRUCTURE OF ATOM

[**For B. Sc. and M. Sc. Students**]

M. G. ARORA
M. Sc, Ph. D.

Anmol Publications Pvt Ltd
New Delhi - 110 002

ANMOL PUBLICATIONS PVT LTD
4374/4B, Ansari Road
Daryaganj, New Delhi-110 002

First Edition 1995

ISBN 81-7488-068-2

PRINTED IN INDIA

Published by J. L. Kumar for Anmol Publications Pvt. Ltd., New Delhi and Printed at Chetna Printers, Maujpur, Delhi-110 053.

Preface

This book deals with all the aspects of the structure of atom. It gives a vivid description of the structure, composition, atomic theories etc. in details. Emphasis has been placed to describe these in simple, illustrative and lucid manner. An effort has been made to include the latest information about the atom and its structure. To explain difficult concepts a large number of illustrations have been included. To make clear certain definitions some numerical problems have been included at appropriate places.

The book is likely to serve as a textbook as well as reference book for the students of Indian Universities.

—Author

Contents

1

Atomic Theory and Atomic Structure

1.1. Introduction

Atomic Theory is central to chemistry. According to this theory, all matter is composed of small particles, or *atoms*. Each element is composed of the same kind of atom, and a *compound* is composed of two or more elements chemically combined in fixed proportions. A *chemical reaction* consists of the rearrangement of the atoms present in the reacting substances to give new chemical combinations present in the substances formed by the reaction. Although Dalton considered atoms to be the ultimate particles of matter, we now know that atoms themselves have structure. An atom has a *nucleus* and *electrons*. These atomic particles were discovered by J.J. Thomson and Ernest Rutherford.

Thomson established that cathode rays consist of negatively charged particles, or electrons, that are constitutents of atoms. Thomson measured the mass-to-charge ratio of the electron, and later Millikan measured its charge. From these measurements, the electron was found to be more than 1800 times lighter than the lightest atom.

Rutherford proposed the nuclear model of the atom to account for the results of experiments in which alpha particles were scattered from metal foils. According to this model, the atom consists of a central core, or nucleus, around which the electrons exist. The nucleus has most of the mass of the atom and consists of *protons* (with a positive charge) and *neutrons* (with no charge). Each chemically distinct atom has a nucleus with a specific number of protons (*atomic number*), and around the nucleus in the neutral atom are an equal number of electron. The number of protons plus neutrons in a nucleus equals the *mass number*. Atoms whose nuclei have the same number of protons but different number of neutrons are called *isotopes.*

The *atomic weight* can be calculated from the isotopic masses and *fractional abundances* of the isotopes in a naturally occurring element. These data can be determined by a mass spectrometer, which separates ions according to their mass to charge ratios.

1.2. Atomic Theory of Matter

Lavoisier laid the experimental foundation of modern chemistry. But it was the British chemist John Dalton (1766-1844) who provided the basic theory : all matter—whether element, compound, or mixture—is composed of small particles called atoms. The postulates, or basic statements, of Dalton's theory are presented in this section.

Postulates of Dalton's Atomic Theory

The main points of Dalton's **atomic theory,** *an explanation of the structure of matter in terms of different combinations of very small particles, are given by the* following postulates :

1. All matter is composed of indivisible atoms. **An atom** is *an extremely small particle of matter that retains its identity during chemical reactions.*

2. An *element* is a type of matter composed of only one kind of atom, each atom of a given kind having the same properties. Mass is one such property. Thus, the atoms of a given element have a characteristic mass.

3. **A compound** *is a type of matter composed of atoms of two or more elements chemically combined in fixed proportions.* The relative numbers of any two kinds of atoms in a compound occur in simple ratios. Water, for example, is a compound of the elements hydrogen and oxygen and consists of hydrogen and oxygen atoms in the ratio of 2 to 1.

4. **A chemical reaction** consist of *the rearrangement of the atoms present in the reacting substances to give new chemical combinations present in the substauces formed by the reaction.* Atoms are not created, destroyed, or broken into smaller particles by any chemical reaction.

Today we know that atoms are not truly indivisible; they are themselves made up of particles, as we will explain in the next section. Nevertheless, Dalton's postulates are essentially correct.

Deductions from Dalton's Atomic Theory

Note how atomic theory explains the difference between an element and a compound. Atomic theory also explains two laws we considered earlier. One of these is the law of conservation of mass, which states that the total mass remains constant during a chemical reaction. By postulate 2, every atom has a definite mass. Because a chemical reaction only rearranges the chemical combinations of atoms (postulate 4), the mass must remain constant. The other law explained by atomic theory is the law of definite proportions (constant composition). Postulate 3 defines a compound as a type of matter containing the atoms of two or more elements is definite proportions. Because the atoms have definite mass, compounds must have the elements in definite proportions by mass.

A good theory should not only explain known facts and laws but also predict new ones. The law of multiple proportions, deduced by Dalton from his atomic theory, is *a law stating that when two elements form more than one compound, the masses of ane element in these compounds for a fixed mass of the other element are in ratios of small whole numbers.* For example, carbon and oxygen form two compounds : carbon monoxide and carbon dioxide. Carbon monoxide contains 1.3321 g of oxygen for each 1.0000 g of carbon, whereas carbon dioxide contains 2.6642 g of oxygen for 1.0000 g of carbon. In other words, carbon dioxide contains twice the mass of oxygen as is contained in carbon monoxide (2-6642 g=2 × 1.3321 g) for a given mass of carbon. Atomic theory explains this by saying that carbon dioxide contains twice as many oxygen atoms for a given number of carbon atoms as does carbon monoxide. The deduction of the law of multiple proportions from atomic theory was important in convincing chemists of the validity of the theory.

Atomic Symbols

We will find it convenient to use symbols for the atoms of the different elements. An atomic symbol is *a one-, -two, or three-letter notation used to represent an atoms corresponding to a particular element.* Typically, the atomic symbol consists of the first letter, capitalized, from the name of the element, sometimes with an additional letter from the name in lower case. For example, chlorine has the symbol Cl. Other symbols are derived from a foreign (usually Latin) name. Sodium is given the symbol Na from its Latin name, natrium. Symbols of selected elements are listed in Table 1.1.

Table 1.1
Some Common Elements

Name of Element	*Atomic Symbol*	*Physical Appearance of Element**
Aluminum	Al	Silvery-white metal
Barium	Ba	Silvery-white metal
Bromine	Br	Reddish brown liquid
Calcium	Ca	Silvery-white metal
Carbon	C	
Graphite		Soft, black solid

Diamond		Hard, colorless crystal
Chlorine	Cl	Greenish-yellow gas
Chromium	Cr	Silvery-white metal
Cobalt	Co	Silvery-white metal
Copper	Cu (from *cuprum*)	Reddish metal
Fluorine	F	Pale yellow gas
Gold	Au (from *aurum*)	Soft, yellow metal
Helium	He	Colorless gas
Hydrogen	H	Colorless gas
Iodine	I	Bluish-black solid
Iron	Fe (from *ferrum*)	Silvery-white metal
Lead	Pb (from *plumbum*)	Bluish-white metal
Magnesium	Mg	Silvery-white metal
Manganese	Mn	Gray-white metal
Mercury	Hg (from *hydrargyrum*)	Silvery-white liquid metal
Neon	Ne	Colorless gas
Nickel	Ni	Silvery white metal
Nitrogen	N	Colorless gas
Oxygen	O	Colorless gas
Phosphorus (white)	P	Yellowish-white, waxy solid
Potassium	K (from *kalium*)	Soft, silvery-white metal
Silicon	Si	Gray, lustrous solid
Silver	Ag from *argentum*)	Silvery-white metal
Sodium	Na (from *natrium*)	Soft, silvery-white metal
Sulfur	S	Yellow solid
Tin	Sn (from *stannum*	Silvery-white metal
Zinc	Zn	Bluish-white metal

1.3. Structure of Atom

Although Dalton had postulated that atoms were indivisible particles, experiments conducted around the beginning of the present century showed that atoms them selves consist of particles. These experiments, which are described in this section, showed that an atom consists of two kinds of particles : a nucleus, *the atom's central core, which is positively charged and contains most of the atom's mass and one or more electron.* An **electron** is *a very light, negatively charged particle that exists in the region around the positively charged nucleus.*

Discovery of the Electron

In 1897 the British physicist J.J. Thomson (Fig. 1) conducted a series of experiments that showed thatatoms were not indivisible particles. Fig. 1.1 shows an experimental apparatus similar to the one used by Thomson. In this apparatus, two electrodes from a high-voltage source are sealed into a glass tube from which the air has been evacuated. The negative electrode is called the *cathode* : the positive one is called the *anode*. When the high-voltage current is turned on, the glass tube emits a greenish light. Experiments showed that this greenish light is caused by the interaction of the glass with *cathode rays,* which are rays that originate from the cathode.

After the cathode rays leave the negative electrode, they move toward the anode, where a portion passes through a hole to form a beam. Although cathode rays are not directly visible, the beam is seen wherever it falls on a luminescent card placed inside the glass tube (the card is coated with a substance that glows, or luminesces, when cathode rays strike it). This beam can be observed to bend according to the voltage applied to the conducting plates placed above and below the luminescent card. In Figure 1.1, the cathode ray beam can be seen to bend away from the negative plate. A similar experiment can be carried out in which cathode rays are seen to bend when a magnet is brought toward them Fig. 1.2. Thomson showed that the characteristics of cathode rays are independent of the material making up the cathode. From such evidence Thomson concluded that a cathode ray consists of a beam of negatively charged particles (or electrons) and that electrons are constituents of all matter.

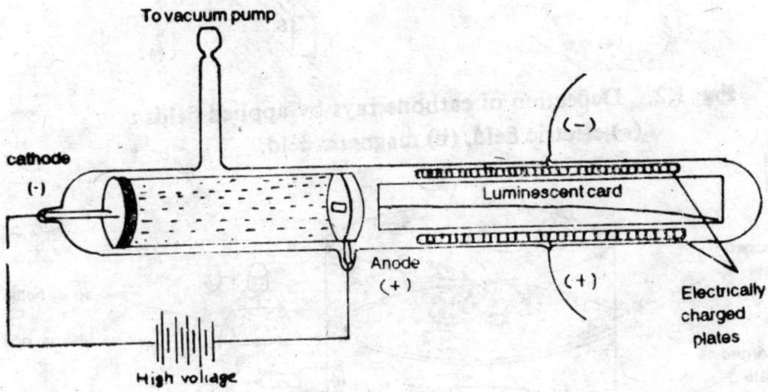

Fig. 1.1 Formation of cathode rays. Cathode rays leave the cathode, or negative electrode, and are accelerated toward the anode, or positive electrode. Some of the rays pass through the hole in the anode to form a beam, which then is bent by the electric plates in the tube. The beam is made visible as a green glow by a luminescent card placed in the tube.

From his experiments, Thomson could also calculate the ratio of the electron's mass, m_e, to its electric charge, e. He could not obtain either the mass or the charge separately, however. In 1909 the U.S. physicist Robert Millikan performed a series of ingenious experiments in which he obtained the charge on the electron by observing how a charged drop of oil falls in the presence and in the absence of an electric field (Figure 1.3). From this type of experiment, the charge on the electron is found to be 1.602×10^{-19} coulombs (the coulomb, abbreviated C, is a unit of electric charge). If you use this charge with the most recent value of the mass-to-charge ratio of the electron, you obtain an electron mass of 9.109×10^{-31} kg, which is more than 1800 times smaller than the mass of the highest atom (hydrogen). This shows quite clearly that the electron is indeed a subatomic particle.

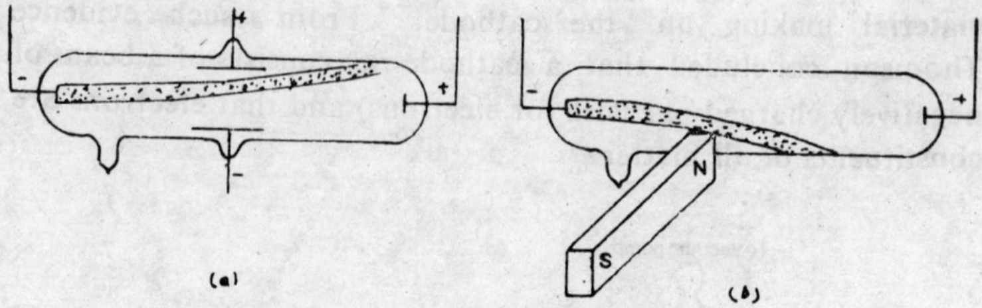

Fig. 1.2. **Deflection of cathode rays by applied fields : (a) electric field, (b) magnetic field.**

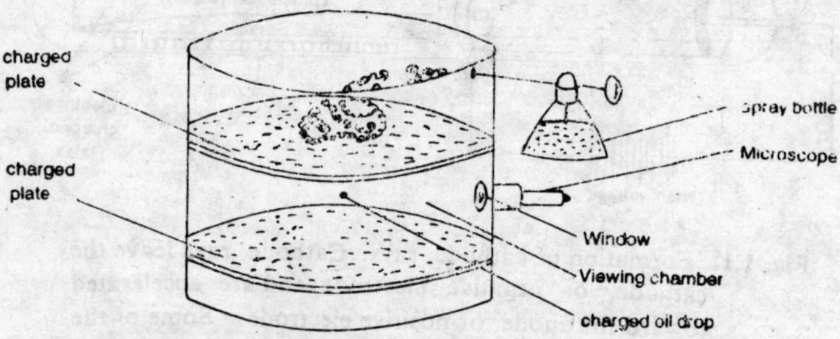

Fig. 1.3. Millikan's oil-drop experiment. An oil drop passes into the region between charged plates, One first observes the drop's speed of free fall, which is related to its mass. If the drop has gained an electron, it can be attracted upward by turning on the current to the electric plates. The drop's upward speed is related to its mass-to-charge ratio from which the charge on the electron can be obtained.

Cathode-ray discharge tube experiments tell us that matter contains negatively charged particles. electrons. Can we learn still more ? Yes, by using a modified discharge tube, one in which the cathode has had a hole drilled in it, as shown in Fig. 1.4. Such a tube was first used by E. Goldstein in 1886. When Goldstein evacuated the tube and applied a high voltage across the electrodes, he observed a glowing streamer emerging from the hole in the cathode on the opposite side from the anode, as is shown in the diagram. Goldstein called this streamer a canal ray. A canal ray has a color depends upon the nature of the residual gas in the tube. Also, because it is possible to deflect the path of a canal ray by using an electric or magnetic field, it can be shown that the ray consists of a mixture of different particles, even if only a single, pure gas is present (at low pressure) in the tube. The particles are all *positively* charged, but if we determine the magnitude of these charges, we find that they are not all the same. Rather, they are multiples of 1.6×10^{-19} C ; that is, we find particles with charge of 1.6×10^{-19} C, 3.2×10^{-19} C, 4.8×10^{-19} C, and so on.

All this can be accounted for as follows : The cathode ray, as we have seen, consists of a stream of electrons which leave the cathode and move toward the anode. Some of these electrons collide with gas molecules and do so vigorously enough to knock out other electrons. Each of these molecules now lacks one or more electrons and consequently carries a *positive charge.* The anode and cathode create an electric field within the discharge tube, and this causes the positively charged molecules to accelerate toward the cathode. Most of them strike the cathode, but a few coast through the hole and can be detected by the glow they produce on the other side.

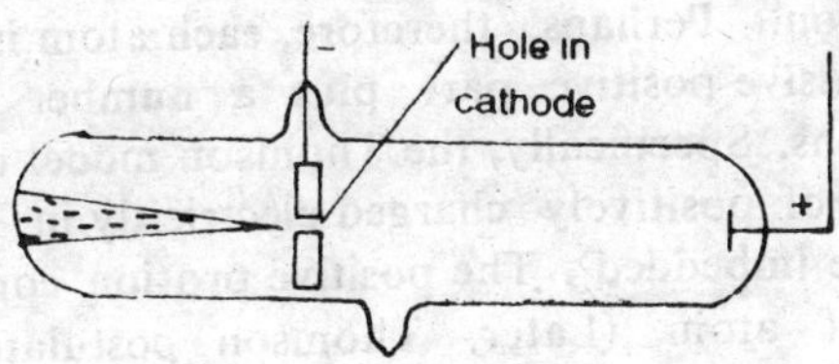

Fig. 1.4. Goldstein's canal-ray tube.

These charged molecules carry various amounts of positive charge depending on how many electrons they have lost. Their masses are always much larger than that of an electron and, in addition, depend upon the identity of the residual gas in the tube.

Why is it true that if an electron, which is *negatively* charged, is removed from a molecule, the molecule acquires a *positive* charge ? If can be shown experimentally that molecules carry no net electrical charge. If removal of an electron causes a molecule to have a positive charge, it must be that molecules are composed of positively charged particles as well as electrons. When the tctal positive charge equals the total negative charge (due to the electrons), the *net* charge is zero. But if one or more electrons are lost, the positive charge now outweighs the negative charge ; so the molecule is said to carry a *net* positive charge.

Atoms or molecules which have gained or lost electrons are called *negative* or *positive ions*, respectively. Each ion is represented by a symbol or formula followed by a superscript indication of the number of unit charges it carries. Thus, Na^+ represents a sodium atoms which *lost* has one electron : S^{2-}, a sulfur atom which has *gained* to electrons ; and O_2^+, an oxygen molecule has lost an electron.

1.4. Thomson Model for Atom

What is present in an atom besides electrons ? What is the atom's structure ? In 1898 J.J. Thomson proposed answer to these questions with a model for the structure of the atom. Thomson reasoned as follows : Electrons can be knocked out of an atom leaving a positive ion which has a much higher mass than an electron. Perhaps, therefore, each atom is composed of a large, massive positive part plus a number of smaller, lighter electrons. Specifically, the Thomson model of the atom was a sphere of positively charged electricitly in which some electrons were imbedded . The positive protion contained most of the mass of atom. (Later, Thomson postulated that the electrons were arranged in rings and were moving in circular orbit through the positively charged sphere.) There was something intuitively satisfying about this model, and it was accepted by physicists and chemists, until the experiments by Rutherford. Geiger, and Marsden, described below, showed that the Thomson model could not be correct.

In 1896 W Rontgen was experimenting with a cathode-ray discharge tube which was totally enclosed by picees of black cardboard. Nearby happened to be lying a piece of paper coated with a compound which, like zinc sulfide, could emit

light when struck by cathode rays. Working in a darkened room, Rontgen noticed that the paper glowed. Later he found that photographic plates (the forerunners of photographic film) in unopened boxes had unexplainably been fogged (exposed). He eventually decided that unknown radiations, or "x rays" had passed through glass, paper, and cardboard, causing the strange phenomenon. X rays are formed when high-energy electrons (cathode rays) strike a metal or other target. They are electromagnetic radiation much like light, but higher in frequency (See 5-3).

Later in the same year the French physicist H. Becquerel showed that compounds of the element uranium would expose a photographic plate, even if the plate were completely protected from light by opaque paper. His work was continued by Marie and Pierre Curic and others who found that several elements released radiations which were similar to x rays. Eventually it was recognized that such "radioactive" substances contain atoms which undergo decay or disintegration. (Atoms are not indestructible). Three kinds of radiation are observed resulting from the disintegration of naturally radioactive elements. These are called α (alpha) particles, β (beta) particles, and γ (gamma) rays. It turns out that α *particles* are rapidly moving nuclei of helium atoms, β *particles* are rapidly moving electrons, and γ *rays* are essentially high-energy x rays.

1.5. Nuclear Model of Atom

Ernest Rutherford (1871 – 1937), a British physicist, put forth the idea of *nuclear model* of the atom in 1911, based on experiments done in his laboratory by Hans Geiger and Ernest Marsden. These scientists observed the effect of bombarding thin gold foil (and other metal foils) with alpha radiation from radioactive substances such as uranium (Fig. 1.5). Rutherford had already showed that alpha rays consist of positively charged particles, Geiger and Marsden found that most of the alpha particles passed through a metal foil as though nothing were there, but a few (about 1 in 8000) were scattered at large angles and sometimes almost backward.

According to Rutherford's model, most of this mass of the atom (99.95% or more) is concentrated in a positively charged center, or nucleus, around which the negatively charged electrons move. Although most of the mass of an atom is in its nucleus, the nucleus occupies only a very small portion of the space of the atom. Nuclei have diameters of about 10^{-15}m (10^{-5} Å), whereas atomic diameters are about 10^{-10}m (1 Å), a hundred thousand times larger. If you were to use a golf ball to represent the nucleus, the atom would be about three miles in diameter.

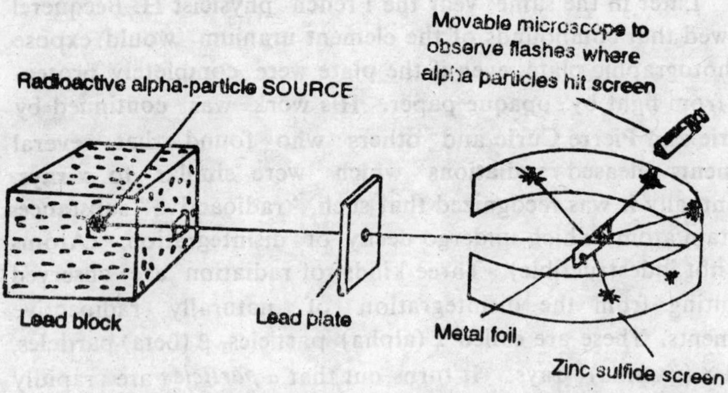

Fig. 1.5. Alpha-particle scattering from metal foils. Alpha radiation is produced by a radioactive source and formed into a beam by a lead plate with a hole in it. (Lead absorbs the radiation.) Scattered alpha particles are made visible by a zinc sulfide screen, which emits tiny flashes where particles strike it. A movable microscope is used for viewing the flashes.

The nuclear model easily explains the results of bombarding gold and other metal foils with alpha particles. Alpha particles are much lighter than gold atoms (alpha particles are helium nuclei). Most of the alpha particles pass through the metal atoms of the foil, undeflected by the lightweight electrons. Whenever an alpha particle does happen to hit a metal-atom nucleus, however, it is scattered at a wide angle because it is repelled by the massive, positively charged nucleus (Figure 1.6).

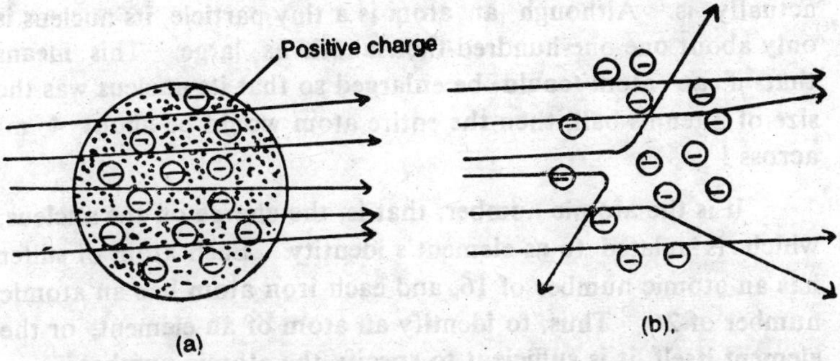

Fig. 1.6. Expected deflections of a particles : (a) Thomson atom, small deflections only; (b) Rutherford atom, large and small deflections.

Rutherford recognized that the Geiger-Marsden results could not be used to distinguish between a positively charged nucleus and a negatively charged one, since a massive negative nucleus would cause large deflections also. He assumed that the nucleus was positive, however, because he knew that electrons have relatively low masses, and if a nucleus consisted of a collection of electrons, then an α particle would be more likely to "flip" the nucleus out of an atom than to be deflected itself. Since deflection of α particles was observed to occur. Rutherford proposed that the *nonelectronic*, that is, *positive* and *massive*, part of the atom was associated with the nucleu.s

Modern Atom

Our present-day model of the atom is a direct descendant of the Rutherford atom. We believe the atom to be composed of two regions : (1) the tiny *nucleus*, composing all the positive charge and practically all the mass of the atom and (2) the *extranuclear region* (everything else), composed of electrons. The positive charge on the nucleus is often expressed in units which are the *magnitude* of the charge on the electron and is called the *atomic number*, Z. For example, an atom of lithium has a nuceus carrying a charge of $+4.8\times10^{-19}$C and three electrons each having a charge of -1.6×10^{-19}C. In electronic charge units, the nucleus of a lithium atom has a 3+ charge, and each of the three electrons a 1− charge. The atomic number (Z) of lithium is 3.

It is important to realize how extremely small the nucleus actually is. Although an atom is a tiny particle, its nucleus is only about one one-hundred-thousandth as large. This means that if an atom could be enlarged so that its nucleus was the size of a tennis ball, then the entire atom would be about 4 mi across !

It is the atomic number, that is, the charge on the nucleus, which is related to as element's identity. Each atom of sulfer has an atomic number of 16, and each iron atom has an atomic number of 26. Thus, to identify an atom of an element, or the element itself, it is sufficient to specify the atomic number.

There is abundant experimental evidence that the nucleus generally consists of two kinds of components particles, or *nucleons*. The first of these is the *proton*, which has a charge (in electronic-charge units) of $+1$. The second of these is the *neutron*, so-named because it is neutral; that is, it carries no charge at all. Since a whole atom carries no net electrical charge, the number of protons in the nucleus must be equal to the number of electrons in the extranuclear region. Also, *the number of protons in the nucleus equals the atomic number*. Except for the ordinary hydrogen nucleus all nuclei contain one or more neutrons. The protons and neutrons in the nucleus of an atom are almost entirely responsible for the atom's mass, because electrons are so very light.

Since the mass of an atom is almost exclusively determined by the total number of nucleons, we call that number the *mass number* and represent it by A. If a particular atom has 47 protons and 60 neutrons, for example, its mass number A is 107.

Any specific atom can be designated using the following shorthand symbolism : Immediately preceding the chemical symbol for the element is written the atomic number Z as a subscript and the mass number A as a superscript. Therefore,

$$ {}^{A}_{Z}X $$

indicates an atom of element X having an atomic number Z and a mass number A*. Since the atomic number is uniquely tied to identity of the element, it is redundant and can be omitted. Thus, for example, we can write either $^{16}_{8}O$, or ^{16}O (each read as "Oxygen 16" or "O 16") to represent this particular atom of oxygen.

All atoms of a given element have the same atomic number and therefore both the same nuclear charge and the same number of electrons in the extranuclear region. Can atoms of a given element differ from each other in any way? Yes, their mass numbers (and hence their masses) can be different. Atoms which have the same atomic number but different mass numbers are called *isotopes*, from Greek words meaning "same place." (If the elements are listed in order of their increasing atomic numbers, all the isotopes of an element occupy the same place in the list). For example, there are three naturally occurring isotopes of oxygen : $^{16}_{8}O$, $^{17}_{8}O$, and $^{18}_{7}O$.

These isotopes differ in mass because different mass numbers mean different numbers of neutrons ($A—Z$), in the nucleus. Thus isotopes are sometimes defined as atoms of the same element having different numbers of neutrons. The numbers of electrons and nucleons in several different atoms are indicated in Table 1.3.

Table 1.2
Numbers of Protons, Neutrons, and Electrons in Various Atoms

Atom	*Population of nucleus*		*Population of extranuclear region*
	Protons	*Neutrons*	*Electrons*
$^{1}_{1}H$	1	0	1
$^{2}_{1}H$	1	1	1
$^{3}_{1}H$	1	2	1

$^{16}_{8}O$	8	8	8
$^{17}_{8}O$	8	9	8
$^{18}_{8}O$	8	10	8
$^{238}_{92}U$	92	146	92

1.6. Nuclear Structure—Isotopes

The nucleus of an atom also has a structure ; it is composed of two different kinds of particles, protons and neutrons. The type of alpha-particle scattering experiment that led to the nuclear model of the atom was also instrumental in clarifying this structure of the nucleus.

An important property to the nucleus is its electric charge. It is possible to deduce the value of the positive charge of a nucleus by analyzing the distribution of alpha particles scattered from a metal foil; other experiments also provide the nuclear charge. From such experiments, researchers discovered that each element has a unique nuclear charge that is an integer multiple of the magnitude of electron charge. This integer, which is characteristic of an element, is called the *atomic number* (*Z*). A hydrogen atom necleus whose magnitude of charge equals that of the electron has the smallest atomic number (1).

In 1911, Rutherford discovered that hydrogen nuclei, or what we now call protons, form when alpha particles strike some of the lighter elements, such as nitrogen. A **proton** is *a nuclear particle having a positive charge equal to* +e *and a mass more than 1800 times that of the electron.* When an alpha particle collides with a nitrogen atom, a proton may be knocked out of the nitrogen necleus. It is the proton in a nucleus that give a nucleus its positive charge.

The **atomic number** (Z) is therefore *the number of protons in the neucleus of an atom*. Because you can determine the

Table 1.3

Properties of the Electron, Proton and Neutron

Particle	*Mass (kg)*	*Charge (C)*	*Mass (amu)**	*Charge (e)*
Electron	9.10939×10^{-31}	-1.60218×10^{-19}	0.00055	-1
Proton	1.67262×10^{-27}	$+1.60218 \times 10^{-19}$	1.00728	$+1$
Neutron	1.67493×10^{-27}	0	1.00866	0

*The atomic mass unit (amu) equals 1.66×10^{-27} kg. it is defined in Section 2.4

atomi- number experimentally, you can determine unambiguously whether or not a sample is a pure element or if perhaps you have discovered a new element. We can now state the definition of an element with more precision. An **element** is *a substance whose atoms all have the same atomic number*. The inside back cover of the book lists the elements and their atomic number (see Table of Atomic Numbers and Atomic Weight.)

The neutron was also discovered by alpha-particle scattering experiments. When beryllium metal is irradiated with alpha rays, a strongly penetrating radiation is obtained from the metal. In 1932 the British physicist James Chadwick (1891-1974) showed that this penetrating radiation consists of neutral particles, or neutrons. The **neutron** is *a nuclear particle having a mass almost identical to that of the proton but no electric charge*. Thus, when beryllium nuclei are struck by alpha particles, neutrons are knocked out. Table 1.2 compares the masses and charges of the electron and the two nuclear particles, the proton and the neutron.

Now consider the nucleus of some atom, say the nucleus of the naturally occurring sodium atom. It contains 11 pertons and 12 neutrons. Thus the charge on the sodium nucleus is $+11e$, which we usually write as simply $+11$, meaning $+11$ untits of electron charge e. Similarly, the nucleus of a naturally occurring aluminum atom contains 13 protons 14 neutrons; the charge on the nucleus is +13.

We characterize a nucleus by its atoms number (Z) and its mass number (A). The mass number (A) is *the total number of protons and neutrons in a nucleus.* The nucleus of the naturally occurring sodium atom has an atomic number of 11, and a mass number of 23(11+12).

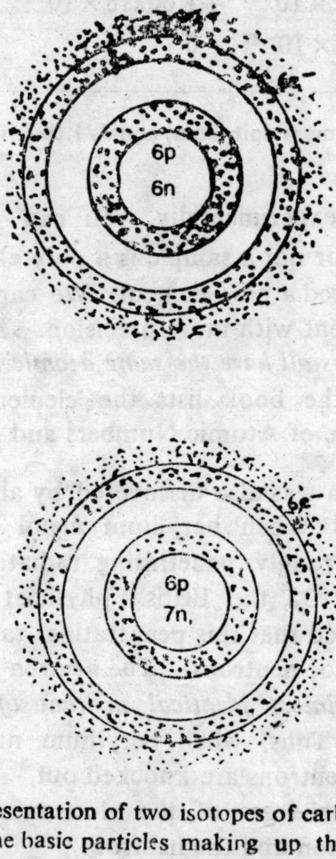

Fig. 1.7. A representation of two isotopes of carbon. The drawing shows the basic particles making up the carbon-12 and carbon-13 isotopes. (The relative sizes of the nuclei are much exaggerated here.)

A **nuclide** is *a particlar nucleus characterized by definite atomic number and mass number.* The shorthand notation for any nucliside consists of the symbol of the element with the atomic number written as a subscript on the left and the mass number as a superscript on the left. You write the *nuclide symbal* for the naturally occurring sodium nuclide as follows :

An atom is normally electrically neutral, so it has as many electrons about its nucleus as the nucleus has protons that is, the number of electrons in a neutral atom equals its atomic number. A sodium atom has a nucleus of charge +11, and around this nucleus there are 11 electrons (with a charge of −11, giving the atom a charge of 0).

A nuclei of the atoms of a particular element have the same atomic number, but the nuclei may have different mass numbers. **Isotopes** are *atoms whose nuclei have the same* ·*tomic number bnt different mass numbers; that is the nuclei have the same number of protons but different numbers of neutrons.* Sodium has only one naturally occurring isotope, which we denote by the same symbol we used for the nuclide (23/11 Na); we also call the isotope sodium-23. Naturally occurring oxygen is a mixture of isotope; it contains 99.759% oxygen-16, 0.037% oxygen-17, and 0.204% oxygen-18. Fig. 1.5 may help you to visualize the relationship among the different subatomic particles in the isotopes of an element.

Ex. 1.1. What is the nuclide symbol for a nucleus that contains 38 protons and 50 neutrons ?

Solution. If you look at the Table of Atomic Numbers and Atomic Weights, you will note that the elements with atomic number 38 (the number of protons in the nucleus) is strontium, symbol Sr. The mass number is 38+50=88.

1.7. Atomic Weights

A central feature of Dalton's atomic theory was the idea that an atom of an element has a characteristic mass. Now we know that a naturally occurring element may be a mixture of isotopes, each isotope having its own characteristic mass. However, the percentages of the different isotopes in most naturally occurring elements have remained essentially constant over time and in most cases are independent of the origin of the element. Thus, what Dalton actually calculated were *average* atomic masses (actually average *relative* masses, as we will discuss). Since you normally work with naturally occurring mixture of elements (whether in pure form or as compounds), it is indeed these average atomic masses that you need in chemical work.

Relative Atomic Masses

Dalton, could not weigh individual atoms. What he could do was to find the average mass of one atom relative to the average mass of another. We will refer to these relative atomic masses as *atomic weights*. To see how Dalton obtained his atomic weights, imagine that you burn hydrogen gas in oxygen. The product of this reaction is water, which is a compound of hydrogen and oxygen. By experiment, you find that 1.0000 g of hydrogen reacts with 7.9367 g of oxygen to form water. To obtain the atomic weight of oxygen (relative to hydrogen), you need to know the relative numbers of hydrogen atoms and oxygen atoms in the water. As you might imagine. Dalton had difficulty with this. Today we know that water contains two atoms of hydrogen for every atom of oxygen. So the atomic weight of oxygen is $2 \times 7.9367 = 15.873$ times that of the mass of the average hydrogen atom.

Atomic Mass Units

Dalton's hydrogen-based atomic weight scale was eventually replaced by a scale based on oxygen and then, in 1961, by the present carbon-12 mass scale, This scale depends on measurements of atomic mass by an instrument called a *mass spectrometer* (Figure 1.8), which we will describe briefly later in this section. You make accurate measurements of mass on this instruments by comparing the mass of an atom to the mass of a particular atom chosen as a standard. On the present atomic weight scale, the carbon-12 isotope is chosen as the standard and is arbitrarily assigned a mass of exactly 12 atomic mass units. One **atomic mass units (amu)** is, therefore, ***a mass unit equal to exactly one-twelfth the mass of a cnrbon-12 atoms.***

On this modern scale, the **atomic weight** of an element is *the average atomic mass for the naturally occurring elements,* *expressed in atomic mass units.* A complete table of atomic weights appears on the inside back cover of this book. (Although the use of the term *weight* in *atomic weight* is not strictly correct, it is sanctioned by convention to mean "average atomic mass.")

Mass Spectrometry and Atomic Weights

Earlier we noted that you use a mass spectrometer to obtain accurate masses of atoms. The earliest mass spectrometers measured the mass-to-charge ratios of positively charged atoms using the same techniques. Thomson used to study electrons (see Figure 1.8). Modern instruments may be other techniques, but all measure the mass-to-charge ratios of positively charged atoms (and molecules as well).

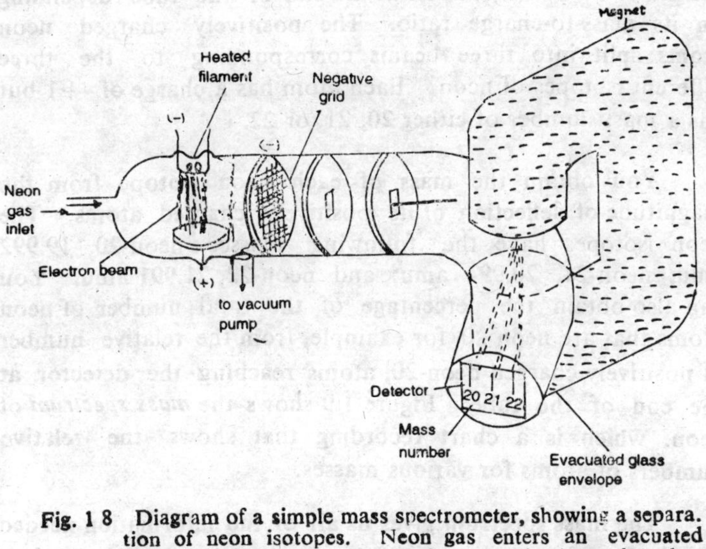

Fig. 1 8 Diagram of a simple mass spectrometer, showing a separation of neon isotopes. Neon gas enters an evacuated chamber, where neon atoms form positive ions when they collide with electrons. Positively charged neon atomr, Ne^+, are accelerated from this region by the negative grid and pass between the poles of a magnet. The beam of positively charged atoms is split into three beams b the magnetic field according to the mass-to-charge ratios. The three beams then travel to a detector at the end of the tube. (The detector is shown here as a photographic plate; in modern spectrometers, the detector is electronic, and the mass positions are recorded on chart paper.)

In the mass spectrometer shown in Figure 1.8, neon gas passes through an inlet tube into a chamber where the neon atoms collide with electron s from an electron beam (cathode rays). The force of a collision can knock an electron from a neon atom. The positive neon atoms produced this way are drawn toward a negative grid, and some of them pass through slits to form a beam of positive electricity. This beam then travels through a magnetic field (from a magnet, whose poles are above and below the positive beam). The magnetic field deflects the positively charged atoms in the positive beam according to their mass-to-charge ratios. Just as a heavy car is buffeted less by a strong wind thad a light car is, the most massive positively charged atoms are deflected the least by the magnetic field. For this reason, each positively charged atom arrives at a definite position at the end of the tube depending on its mass-to-charge ratio. The positively charged neon atoms split into three beams corresponding to the three different isotopes of neon. Each atom has a charge of +1 but has a mass number of either 20, 21, or 22.

You obtain the mass of each neon isotope from the magnitude of deflection of its positively charged atoms. The neon isotopes have the following masses : neon-20, 19.992 amu; neon-21, 20.994 amu; and neon-22, 21.991 amu. You can also obtain the percentage of the total number of neon atoms that are neon 20. for example, from the relative number of positively charged neon-20, atoms reaching the detector at the end of the tube. Figure 1.9 shows the *mass spectrum* of neon, which is a chart recording that shows the relative numbers of atoms for various masses.

The mass spectrum gives us all of the information needed to calculate the atomic weight; the masses of isotopes and their relative numbers, or fractional abundances. The **fractional abundance** of an isotope is *the fraction of the total number of atoms that is composed of a particular isotope.* The fractional abundances of the neon isotopes in naturally occurring neon are neon-20, 0.9051; neon-21, 0.0027; and neon-22, 0.0922.

You calculate the atomic weight of an element by multiplying each isotopic mass by its fractional abundance and summing the values. If you do that neon using the data given here, you will obtain 20.179 amu. The next example illustrates the calculation in full for chromium.

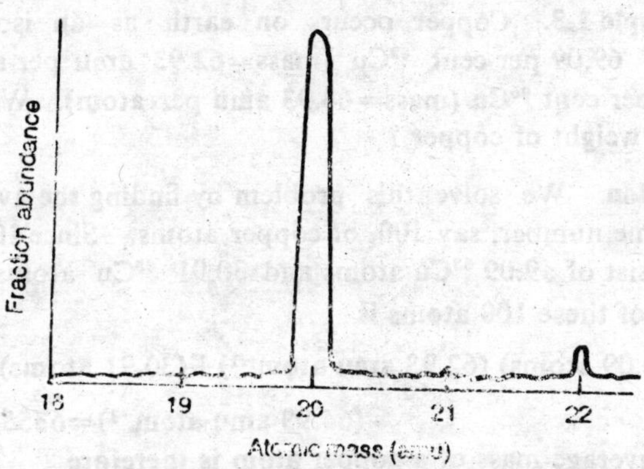

Fig. 1.9. The mass spectrum of neon. Neon is separated into its isotopes Ne-20, Ne-21 and Ne-22. The height at each mass peak is proportional to the fractional of the fraction of that isotope in the element.

Example 1.2 Chromium, Cr. has the following isotopic masses and fractional abundances :

Mass Number	*Isotopic Mass (amu)*	*Fractional Abundance*
50	49.9461	0.0435
52	51.9405	0.8379
53	52.9407	0.0950
54	53.9389	0.0236

What is the atomic weight of chromium ?

Solution. Multiply each of the isotopic masses by its fractional abundance, then sum :

$$49.9461 \text{ amu} \times 0.0435 = 2.17 \text{ amu}$$

$$51.9405 \text{ amu} \times 0.8379 = 43.52 \text{ amu}$$

$$52.9407 \text{ amu} \times 0.0950 = 5.03 \text{ amu}$$
$$53.9389 \text{ amu} \times 0.0236 = 1.27 \text{ amu}$$
$$\overline{}$$
$$51.99 \text{ amu}$$

The atomic weight of chromium is 51.99 amu.

Example 1.3. Copper occurs on earth as an isotopic mixture of 69.09 per cent ^{63}Cu (mass=62.93 amu per atom) and 30.91 per cent ^{65}Cu (mass=64.93 amu per atom). What is the atomic weight of copper ?

Solution. We solve this problem by finding the average mass of some number, say 100, of copper atoms. Since 100 Cu atoms consist of a9.09 ^{63}Cu atoms and 30.91 ^{65}Cu atoms, the total mass of these 100 atoms is

$$(69.09 \text{ atoms})(62.93 \text{ amu atom}^{-1}) + (30.91 \text{ atoms})(64.93 \text{ amu atom}^{-1}) = 6355 \text{ amu.}$$

The average mass of a copper atom is therefore

$$\frac{6355 \text{ amu}}{100 \text{ atoms}} = 63.55 \text{ amu atom}^{-1}$$

It is important to remember that the atomic weight scale is a *relative* scale. The atomic weight of an element is the average mass of the atoms of that element expressed relative to the mass of a 12/6C atom, which is set equal to 12. Consider, for instance, the atomic weight of magnesium, which is known to be 24.3. This means that one "average") magnesium atom has a mass of 24.3 amu. Since the mass of a 12/6C atom is 12 amu, one Mg atom weighs a little more than twice as much as one C atom. Similarly, chlorine, atomic weight 35.5 weighs a little less than three times as much as a 12/6C atom.

2

Quantum Theory of Atom

2.1. Introduction

One way to study the electronic structure of the atom is to analyze the electromagnetic radiation that is emitted from an atom. Electromagnetic radiation is characterized by its *wavelength* λ and frequency ν, and these quantities are related to the speed of light c ($c=\nu\lambda$).

Light was shown by Einstein to consist of particles (*photons*), each of energy $E=h\nu$, where h is Planck's constant. According to Bohr, electrons in an atom have *energy levels*, and when an electron in a higher energy level drops (or undergoes a transition) to a lower energy level, a photon is emitted. The energy of the photon equals the difference in energy between the two levels.

Fireworks displays are fascinating. The spectacular colors of fireworks are due to the presence of metal compounds in the burning material. Strontium nitrate, carbonate, or sulfate gives a deep red color; barium nitrate or chloride a green, and copper nitrate or sulfate a bluish-green.

Chemists began studying colored flames in the eighteenth century and soon used *flame tests* to distinguish sodium from potassium compounds. A flame test is a simple way to identify an element. When sodium compounds are sprayed into a flame, the flame burns with a bright yellow color. Potassium gives a violet flame, and lithium and strontium give a red.

Although the red flames of lithium and strontium appear similar, the light from each can be resolved (or separated) by means of a prism into distinctly different colors. The resolution easily distinguishes the two elements. A prism disperses the colors of white light just as small raindrops spread the colors of sunlight into a rainbow or spectrum. But the light from a flame, when passed through a prism, reveals something other than a rainbow. Instead of the continuous range of color from red to yellow to violet, the spectrum from a strontium flame, for example, shows a cluster of red lines and blue lines against a black background. The spectrum of lithium is different, showing a red line, a yellow line, and two blue lines against a black background.

Each element, in fact, has a characteristic line spectrum because of the emission of light from atoms in the hot gas. These spectra can be used to identify elements. Looking at such spectra, you can ask many questions about their meaning. How is it that each atom emits particular colors of light ? What does a line spectrum tell you about the structure of an atom ? If you know something about the structures of atoms, can you explain the formation of irons and molecules ? We will answer these questions in this and the next few chapters.

2.2. Quantum Effects and Photons

Isaac Newton, who studied the properties of light in the seventeeth century, believed that light consisted of a beam of particles. In 1801, however, British physicist Thomas Young showed that light, like waves, could be diffracted. *Diffractions* is a property of waves in which the waves spread out when they encounter an obstruction or small hot about the size of the wavelength. You can observe diffraction by viewing a light source through a hole—for example, a street light through a mesh curtain. The image of the street light is blurred by diffraction.

By the early part of this century, the wave theory of light appeared to be well entrenched. But in 1905 the German-born U.S. physicist Albert Einstein (1879—1955) discovered that he could explain a phenomenon known as the *photoelectric effect* postulating that light had both wave and particle properties. Einstein based this idea on the work of the German physicist Max Planck (1858-1947). In this section, we will took at Planck's work and at Einstein's explanation of the photoelectric effect. We also discuss compton effect.

2 3. Planck's Quantization of Energy

In 1900 Max Planck found a theoretical formula that exactly describes the intensity of light of various frequencies emitted by a hot solid at different temperatures. Earlier, others had shown experimentally that the light of maximum intensity from a hot solid varies in a definite way with temperature. A solid glows red at 750°C, then white as the temperature increases to 1200°C. At the lower temperature, chiefly red light is emitted. As the temperature increases, more yellow and blue light become mixed with the red, giving white light.

According to Planck, the atoms of the solid oscillate, or vibrate, with a definite frequency ν. But in order to reproduce the results of experiments on glowing solids, he found it necessary to accept a strange idea. A vibrating atom could have only certain energies E, those allowed by the formula.

$$E = nh\nu, \; n = 1.\ 2,\ 3 \ldots$$

where h is a constant, now called **Plank's constants,** *a physical constant with the value* 6.53×10^{-34} *J.s.* The value of n must be 1 or 2 or some other whole number. Thus, the only energies a vibrating atom can have are $h\nu$, $2h\nu$, $3h\nu$, and so forth.

The numbers symbolized by n are called *quantum numbers.* The vibrational energies of the atoms are said to be *quantized*; that is, the possible energies are limited to certain values. Quantization of energy seems contradicted by everyday experience. Imagine the same concept being applied to the energy of a moving automobile. Quantization of a car's energy would mean that only certain speeds were possible. A car could travel at, say, 10, 20, or 30 miles per hour (mi/h) but not at 12 or 25 mi/h or at any other intermediate speeds. For car energies, this seems unreasonable. But for atoms, quantization is the rule.

In the study of the physical world we find two general kinds of physical quantities : quantities that have a continuum of values and quantities that have only discrete. or *quantized*, values.

Figure 2.1 shows several examples of *classical* continuous, or non-quantized, physical quantities.

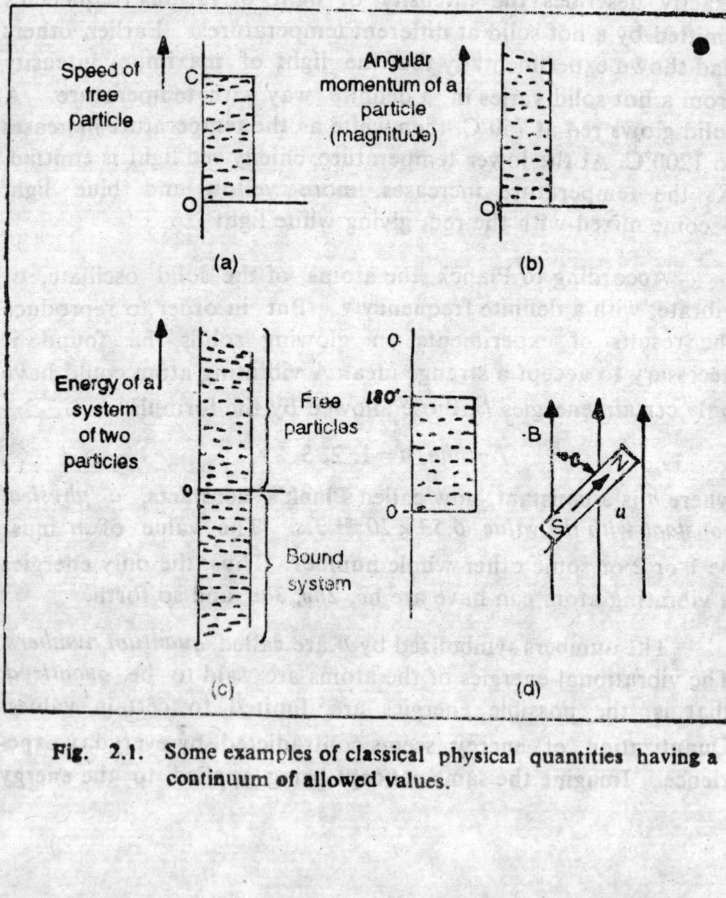

Fig. 2.1. Some examples of classical physical quantities having a continuum of allowed values.

(a) The speed of a free particle, which can range continuously from zero to the speed of light.

(b) The magnitude of the augular momentum of a particle, which can have any value from zero to infinity.

(c) The mechanical energy of a system of two particles, which can assume any negative value when the particles are bound together ($E_m<0$) and any positive value when the particles are free ($E_m>0$).

(d) The angle between the direction of a magnet's dipole moment μ and an external magnetic field B, which may range continuously from 0 to 180°.

Figure 2.2 shows several examples of quantized physical quantities :

(a) The speed of light in a vacuum, always *c*.

(b) Electric charge, which is quantized in that the total charge of any body is precisely an integral multiple, either positive or negative, of the fundamental electron charge *e*. The quantization of charge, clearly revealed in the chemical idea of valence and in the laws of electrolysis, was most directly demonstrated in the oil-drop experiments of R.A. Milliken (1910), in which the charge of the electron was measured directly.

(c) Standing waves and resonance, which are particularly striking manifestations of quantization in classical physics. The frequency of oscillation of a resonating vibrating string, fixed at both ends, can be only an integral multiple of the lowest, or fundamental, frequency of oscillation f_0. The fundamental frequency is determined in turn by the physical properties of the string and its length. The wave on the string is repeatedly reflected from the boundaries, or fixed ends, and constructively interferes with itself, so to speak, to produce standing waves. Resonance can be achieved only if the distance between the end points is precisely an integral multiple of half-wavelengths. It is known that the frequency of a wave is precisely determined only when the wave has an infinite extension in space; this argument is valid ever for a wave trapped between reflecting boundaries, because it can be imagined that the wave is folded on itself an infinite number of times.

(d) The observed rest masses of atoms, which do not occur in a continuous range. This was first perceived in the fundamental studies of chemical combination that led to the atomic theory of Dalton. The masses of atoms occurring in nature are now known very precisely, and whereas they are nearly in the ratio of integers, they are not precisely so. One of the principal tasks of nuclear science is to explain on some fundamental basis these departure from integral ratios.

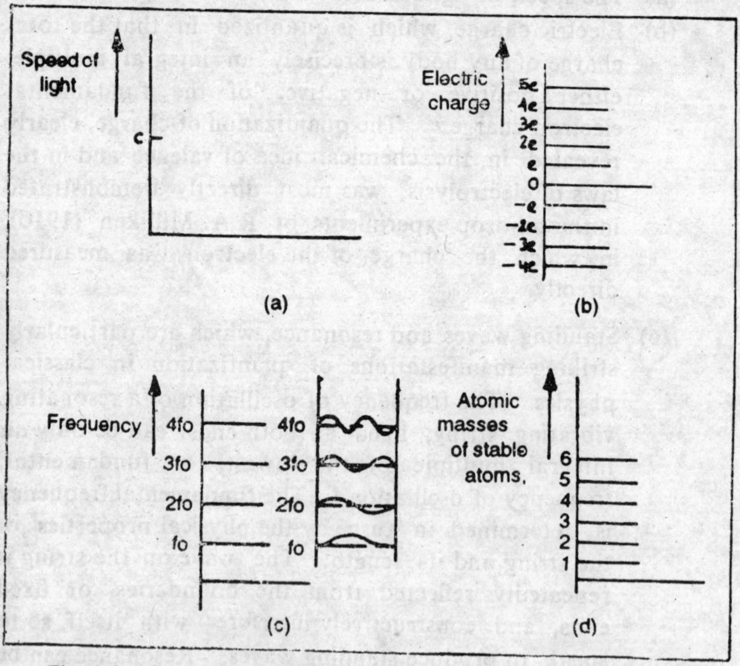

Fig. 2.2. Some examples of classical physical quantities having quantized values.

The quantum theory is largely based on the discovery that certain quantities hitherto regarded in classical physics as continuous are, in fact, quantized. Historically the quantum theory originated in the theoretical interpretation of electromagnetic radiation from a blackbody (a perfect absorber and radiator). Near the end of the 1800s, it was found that the experimentally observed variation with wavelength of the intensity of electromagnetic radiation from a blackbody disagreed with the theoretical expectations of classical electromagnetism. Max Planck, formulator of the quantum theory, showed in 1900 that a revision of classical ideas, through the concept of energy quantization, led to satisfactory agreement between experiment and theory. Because a detailed analysis of blackbody radiation involves sophisticated arguments, we shall introduce the quantum concepts through the simpler and more compelling considerations that arise in the photoelectric effect.

2.4. Photoelectric Effect

Planck himself was uneasy with the quantization assumption and tried (unsuccessfully) to eliminate it from his theory. Albert Einstein, on the other hand, boldly extended Planck's work to include the structure of light itself. Einstein reasoned that if a vibrating atom changed energy, say from $3h\nu$ to $2h\nu$, it would decrease in energy by $h\nu$, and this energy would be emitted as a bit (or quantum) of light energy. He therefore postulated that light consists of quanta (now called photons), *or particles of electromagnetic energy, while energy E proportional to the observed frequency of the light.*

$$E = h\nu$$

In 1905 Einstein used this photon concept to explain the photoelectric effect.

The *photoelectric effect is the ejection of electrons from the surface of a metal or from another material when light shine on it* (see Figure 2.3). Electrons are ejected, however, only when the frequency of light exceeds a certain *threshold value* characteristics of the particular metal. For example, although violet light will cause potassium metal to eject electrons, no amount of red light (which has a lower frequency) has any effect.

To explain this dependence of the photoelectric effect on the frequency, Einstein assumed that an electron is ejected from a metal when it is struck by a single photon. Therefore, this photon must have at least enough energy to remove the electron from the attractive forces of the metal. No matter how many photons strike the metal, if no single one has sufficient energy, an electron cannot be ejected. A photon of red light has insufficient energy to remove an electron from potassium. But a photon corresponding to the threshold frequency has just enough energy, and at higher frequencies it has more than enough energy. When the photon hits the metal, its energy $h\nu$ is taken up by the electron. The photon ceases to exist as a particle; it is said to be absorbed.

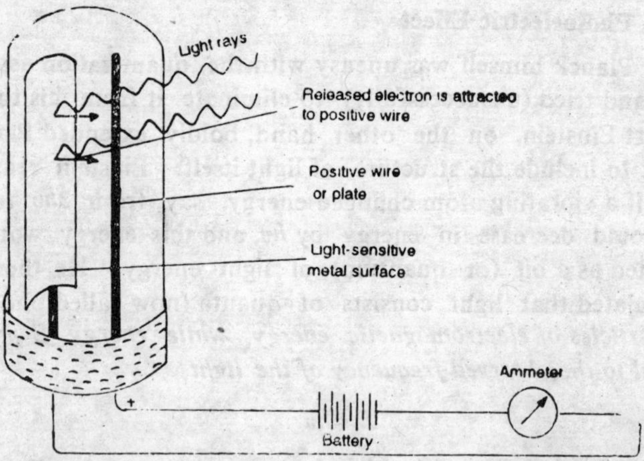

Fig. 2.3. The photoelectric effect. Light shines on a metal surface, knocking out electrons. The metal surface is contained in an evacuated tube, which allows the ejected electrons to be accelerated to a positively charged plate. As long as light of sufficient frequency shines on the metal, free electrons are produced and a current flows through the tube. (The current is measured by an ammeter.) When the light is turned off, the current stops flowing.

The wave and particle pictures of light should be regarded as complementary views of the same physical entity. **This is called the wave-particle duality of light.** Neither the wave nor the particle view alone is a complete description of light.

Experimental Results of the Photoelectric Effect

The results of experiments on the photoelectric effect are summarized in Figure 2.4. We shall take them up in the order in which they are given in the figure.

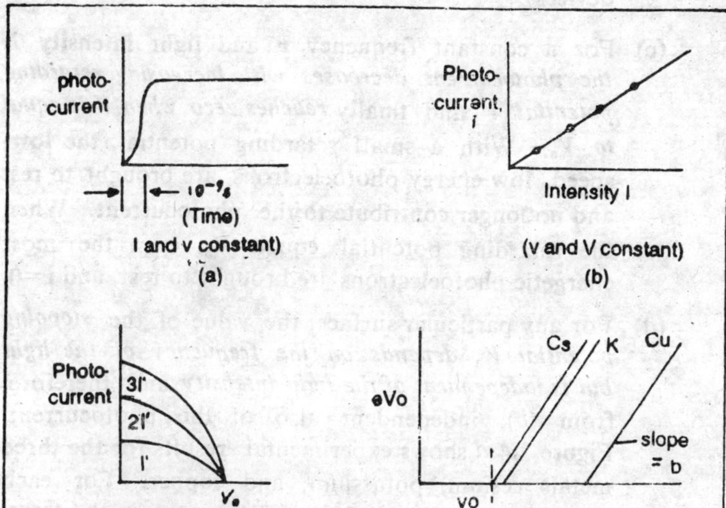

Fig. 2.4. Experimental results of photoelectric emission

(a) When light shines on a metal surface and photoelectrons are emitted, the photocurrent begins almost instaneously, even when the light beam has an intensity as low as 10^{-10} W/m^2 (the intensity at a distance of 200 ml from a 100-W light source). The delay in time, from the instant that the light beam first shines on the surface until photoelectrons are first emitted, is no larger than 10^{-9} s.

(b) For any fixed frequency and retarding potential, the *photocurrent i* is directly *proportional to the intensity I* of the light beam. Since the photocurrent is a measure of the number of photoelectrons released per unit time at the anode and collected at the cathode the relation signifies that the number of photoelectrons emerging per unit time is proportional to the light intensity. (The variation in photocurrent with intensity is used in practical photoelectric devices).

(c) For a constant frequency v and light intensity I, *the photocurrent decreases with increasing retarding potential V* and finally *reaches zero when V is equal to V_0*. With a small retarding potential, the low-speed, low-energy photoelectrons are brought to rest and no longer contribute to the photocurrent. When the retarding potential equals V_0, even the most energetic photoelectrons are brought to rest, and **i=0**.

(d) For any particular surface, the value of the *stopping potential V_0 depends on the frequency of the light but is independent of the light intensity* and therefore. from (*b*), independent also of the photocurrent. Figure 2.4 *d* shows experimental results for the three metals cesium, potassium, and copper. For each metal there is a well-defined frequency v_0, the *threshold* frequency, which must be exceeded to photoemission to occur at all; that is, no photoelectrons are produced, however great the light intensity unless $v > v_0$. For most metals the threshold freuency, lies in the region of ultraviolet light, and a typical stopping potential is several volts. The emitted photoelectrons have energies of several electron volts; therefore, we are justified in using the classical kinetic-energy formula for the photoelectrons.

For any one type of metal the experimental results of Figure 2.4 *d* can be represented by the straight-line equation

$$eV_0 = hv - hv_0 \qquad \ldots(1)$$

where h, representing the slope of the straight line, is found to be the *same* for *all* metals, and ν_0 is the threshold frequency for the particular metal. Rearranging terms and using Equation 4–1 gives

$$h\nu = \frac{1}{2} m_0 v^2_{max} + h\nu_0 \qquad ...(2)$$

Since $\frac{1}{2} m_0 v^2_{max}$ has the dimensions of energy, the terms $h\nu$ and $h\nu_0$ must also have the dimensions of energy.

Classical Interpretation of the Photoelectric Effect

Certain effects can be expected on the basis of the classical properties of electromagnetic waves for each of the four experimental results on the photoelectric effect just described. We consider them with reference to Figure 2.4 and in the same order.

(a) Because of the apparently continuous nature of light waves, the energy absorbed on the photoelectric surface should be proportional to the intensity of the light beam (the power per unit area), the area illuminated, and the time of illumination. All electrons that are bound to the surface of the metal with the same energy must be regarded as equivalent, and any one electron will be free to leave the surface only after the light beam has been on long enough to supply the electrons's binding energy. Moreover, since any one electron is equivalent to any other electron bound with the same energy, when one electron has accumulated enough energy to be freed, numerous other electrons shouid have, too. Independent experiments show that in a typical metal, the least energy with whieh an electron is bound to the surface is a few electron volts. A conservative calculation shows that in the case of an intensity as low as 10^{10} W/m², for which delay times no larger than 10^{-9} s have been observed, no photomission can be expected until at least several hundred hours have elapsed ! Clearly, the classical theory can't account for the essentially instaneous photoelectric emission.

(b) Classical theory predicts that as the light intensity is increased, so is the energy absorbed by electrons at the surface. Hence, the number of photoelectrons emitted, or the photocurrent, is expected to increase proportionately with the light intensity. Here classical theory agrees with the experimental result.

(c) The results of these observations show that there is a distribution in the speeds, or energies, of the emitted photoelectrons; the distribution is in itself not incompatible with classical theory, because it can be attributed to the varying degrees of binding of electrons at the surface or to the varying amounts of energy extracted by electrons from the incident light beam. That there is, however, a very well-defined stopping potential V_0 for a given frequency, independent of the intensity, indicates that the maximum energy of released electrons in no way depends on the total amount of energy reaching the surface per unit time. Classical theory predicts no such effect.

(d) The existence of a threshold frequency for a given metal, a frequency below which no photoemission occurs however great the light intensity, is completely inexplicable in classical terms. In the classical view, the primacy circumstances that determines whether or not photoemission will occur is the energy reaching the surface per unit time (or the intensity), and not the frequency. Further, the appearance of a single constant *h* that relates through Equation 4–2, the maximum energy of photoelectrons to the frequency for any material cannot be understood in terms of any constants of classical electromagnetism.

In short, *classical electromagnetism cannot give a reasonable basis for understanding the experimental results illustrated in Figure, 2.4 a. c, and d.*

Quantum Interpretation of the Photoelectric Effect

An understanding of the photoelectric effect is to be found through the quantum theory. Albert Einstein first applied the quantum theory to electromagnetic radiation in 1905; this led to a satisfactory explanation of photoelectric effect experiments that R.A. Milikan carried out in 1916.

According to the quantum theory, the apparently continuous electromagnetic waves are quantized and consist of discrete *quanta*, called *photons*. Each photon has an energy E that depends only on the frequency (or on the wavelength) and is given by

$$E = hv = h\frac{c}{\lambda} \qquad ...(3)$$

The constant h is the very same h that appears in Equation 4—2, which summarizes the results of experiments on the photoelectric effect. This fundamental constant of the quantum theory is called *Planck's constant* because its value was first determined and its significance first appreciated by Planck in 1900 in interpreting blackbody radiation. The present value of Planck's constant is found by experiment to be

$$h = 6.6262 \times 10^{-34} \text{ J.s}$$

From the quantum point of view, a beam of light of frequency v consists of particlelike photons, each with energy hv. A single photon can interact only with a single electron at the metal surface of a photoemitter: it cannot share its energy among several electrons. Since a photon travels with the speed of light, it must, by relativity theory, have zero rest mass and an energy that is then entirely kinetic. When a particle with a zero rest mass ceases to move with a speed c, it ceases to exist; so long as it exists, it moves at the speed of light. Thus, when a photon strikes an electron bound in a metal and no longer moves at the unique speed c, it relinquishes its entire energy $h\nu$ to the single electron it strikes. If the energy the bound electron gains from the photon exceeds the energy binding it to the metal surface, the excess energy becomes the kinetic energy of the photoelectron.

By the quantum theory, we are prepared to interpret the experimental results of the photoelectric effect. We examine them in reverse order for convenience, referring to Figure 2.4.

The terms in Equation now give a simple meaning to the energies of the photon and photoelectron.

$$h\nu = 1/2m_0 v^2_{max} + h\nu_0 \quad (4)$$

The left side of this equation gives the energy carried by a photon and supplied to a bound electron. The electrons that are least tightly bound leave the surface with maximum kinetic energy. The right side of Equation gives the energy the electron gains from the photon, namely, the kinetic energy and the binding energy. The binding energy of the electrons least tightly bound to the metal surface, often represented by φ, is called the *work function* ; it represents the work needed to remove the least tightly bound electron. Therefore,

$$\varphi = h\nu_0 \quad ...(5)$$

and Equation 2 can be written in the form

$$h\nu = 1/2m_0 v^2_{max} + \varphi \quad ...(6)$$

The value of φ for a particular material, determined from the photoelectric effect, agrees with the value of the work function obtained through independent experiments based on different physical principles. An electron bound with an energy φ can be released only if a single photon supplies at least this much energy, that is, only if $h\nu > \varphi = h\nu_0$ or $\nu > \nu_0$. Figure 2-4 *d* then takes on new meaning. The ordinate can now be identified with the photon energy, as shown in Figure 4-5. (Figure 4-5 should also be compared with the right-hand side of Figure 3-8, where for the photoelectric effect, $E_b = \varphi$).

(a) A well-defined maximum kinetic energy of photoelectrons exists for any given frequency, because the frequency of the electromagnetic radiation determines precisely the photon energy ($E = h\nu$),

(b) The intensity of a monochromatic electromagnetic wave takes on a new meaning. It is, in the quantum view, the energy of each photon multiplied by the number of photons crossing a transverse unit area per unit time. An increase in the intensity of a light beam means, therefore, a proportionate increase in the number of photons striking the metal surface. It is then expected that the number of photoelectrons or the photocurrent i will be proportional to I.

(c) Photoemission occurs with no appreciable delay, because whether an electron is released depends, even at the smallest intensity, not on its accumulating energy but simply on its being hit by a photon, which on stopping relinquishes all of its energy to the electron.

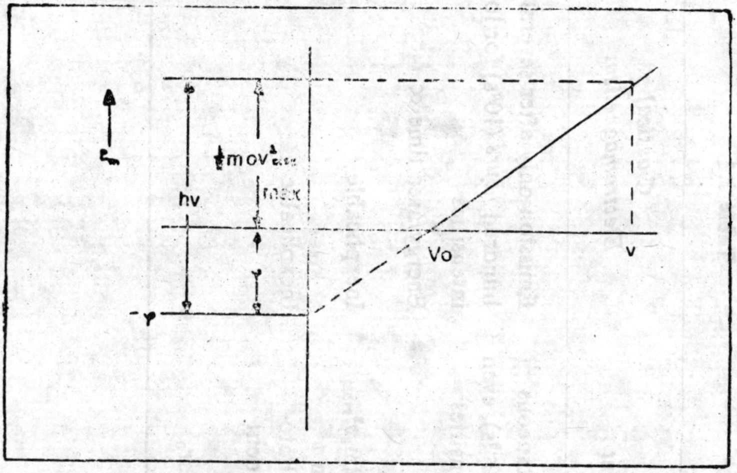

Fig. 2.5. Maximum kinetic energy of photoelectrons as a function of the frequency of the incident photons for a particular material.

Table 2.1 summarizes the results of experiment, the classical interpretation and the quantum interpretation, for each of the four effects shown in Figure 2.4.

Our discussion of the photoelectric effect has thus far concerned the effects found when visible or ultraviolet light shines on a *metal* surface. The first detailed experiments that

Table 2.1

Effect (Figure 2.4)	*Experiment*	*Classical Electromagnetism*	*Quantum Theory*
a	Essentially instantaneous photoemission (10^{-9}s), even for every low intensities	Emission only after several hundred *hours* (10^6s) for low intensities	A single photon gives its energy to a single electron essentially instantaneously
b	$I \propto i$	Energy/area time $\propto i$	$I \propto$ number of photons $\propto i$/area-time.
c	A well-defined $1/2 m_0 v^2_{max}$ dependent only on v	Inexplicable	A photon gives all its energy to a single electron
d	A threshold for photoemission, independent of I and i : $h_{\alpha} = 1/2 m_0 v^2_{max} + hr_0$	Inexplicable	Photon energy$=hv$; work function$=\varphi=hv_0$

led historically to Einstein's quantum interpretation were performed with metal surfaces. The effect occurs on materials other than metals and with photons of different frequencies and energies, however. The photoelectric effect can occur whenever a photon strikes a *bound* electron with enough energy to exceed the binding energy of the electron. An example is a photon freeing a bound electron from a single atom. The phenomenon is one of the most important interactions between short-wavelength electromagnetic radiation and atoms. When a high-frequency (high energy) photon, such as an x-ray or γ-ray, strikes an atom, an electron bound with an energy E_b can be released, provided that $hv > E_b$; then the photoemission results in atomic ionization. The kinetie energy of the released photoelectron generally must be written in the relativistic form $E - E_0$, and so the general form of Equation 2, the energy equation of the photoelectric effect, becomes

$$hv = (E - E_0) + E_b \qquad \ldots(7)$$

The photoelectric effect thus provides an indirect method of measuring the energy of a photon. Suppose that the photoelectron's kinetic energy $E - E_0$ is measured and the binding energy E_b is known on some other basis; then hv can be computed from Equation 7. Conversely, if hv and $E - E_0$ are measured, then E_b can be determined.

The photoelectric effect is only one of several ways in which protons can be removed from a beam of electromagnetic radiation. The photoelectric effect can occur simultaneously with the competitive processes of the Compton effect and pair production.

The new and fundamental insight into electromagnetic radiation that the photoelectric effect provides is the quantization of electromagnetic waves, or the existence of photons. We can properly speak of the quantization of electromagnetic waves since the radiation can be regarded as a collection of particlelike photons, each of energy hv. When the frequency of the radiation is specified as v, the photon can have but one energy, hv. The total energy of a beam of monochromatic electromagnetic radiation is always precisely an integral multiple of the energy $h\nu$ of a single photon (Figure 2.6).

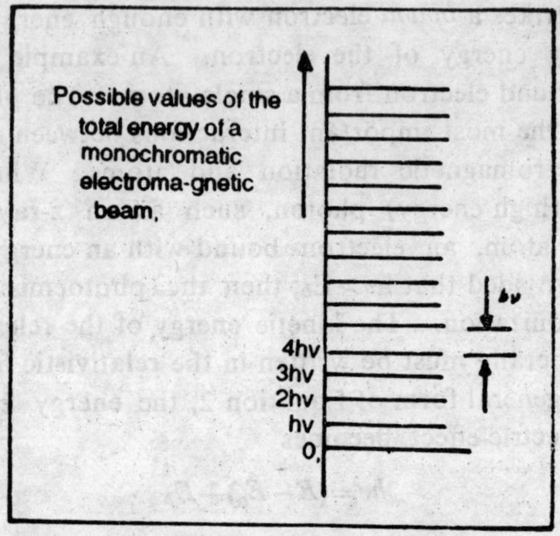

Fig. 2.6. Allowed energies of a beam of monochromatic electromagnetic radiation.

The granularity of electromagnetic radiation is not conspicuous in ordinary observations since the energy of any one photon is very small and the number of photons in a light beam of moderate intensity is enormous. The situation is like that found in the molecular theory—the molecules are so small and t eir numbers so great that the molecular structure of all matters is disclosed only in very subtle observations.

The electromagnetic spectrum, often scaled in units of frequency, can be scaled by the quantum theory in units of energy per photon (see Figure 2.7).

Electromagnetic waves are commonly characterized by their wavelength. It is useful to have a relation between the photon energy in electron volts and the corresponding wavelength in angstroms. For example, x-radiation of wavelength 1.0 angstrom (abbreviated Å) consists of photons whose energy E is

$$E = h\nu = \frac{hc}{\lambda} = \frac{(6.626\times10^{-34}\ \text{J.s})\,(2.998\times10^{8}\ \text{m/s})}{(1.0\ \text{Å})\,(10^{-10}\ \text{m/Å})\,(1.602\times10^{-19}\ \text{J/eV})}$$

$$= 1.240\times10^{4}\ \text{eV} = 12.40\ \text{keV}$$

More generally, the photon energy E, in electron volts, can be written in terms of the wavelength λ in angstroms :

$$E = \frac{1.240\times10^{4}\ \text{eV.Å}}{\lambda} = \frac{0.01240\ \text{MeV.Å}}{\lambda} \tag{8}$$

The energy per photon is smallest for radio-wave photons ($\sim10^{-12}$ eV) and largest for γ ray photons ($\sim$1 GeV). The electromagnetic frequency spectrum corresponds exactly to the energy spectrum of a zero rest-mass particle, or photon, whose energy can extend from zero to infinity, as shown in Figure 2.7. The rest energies of the electron and proton are also shown for comparison.

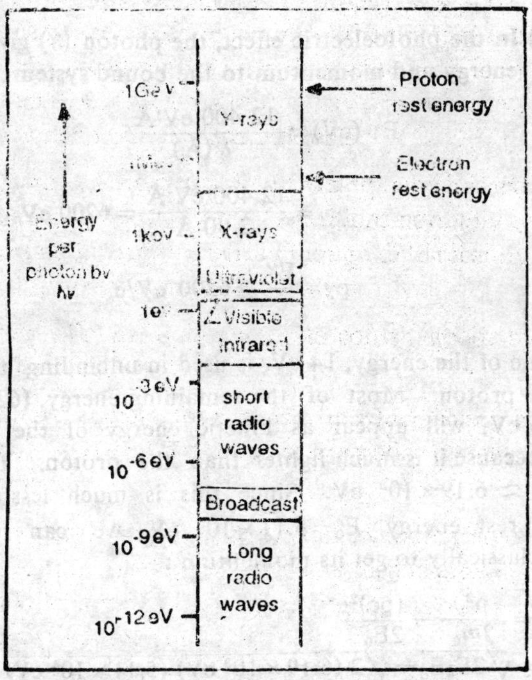

Fig. 2.7. Spectrum of electromagnetic radiation on scale of photon energy.

As we saw earlier that the ideas of wave and particle seem mutually incompatible, even contradictory. That the photoelectric effect shows light behaving as if it consisted of particles or photons does not mean that we can dismiss the incontrovertible experimental evidence of the wave properties of light. Both descriptions must be accepted. An account of the way in which this dilemma is resolved is postponed (Section 5.4) until after we have explored more fully the quantum attributes of light.

Example 2.1. A 2.00-Å photon interacts with a bound electron in the hydrogen atom (binding energy, 13.6 eV). If the photoelectric effect occurs and the electron moves forward in the same direction as the incident photon, what are the energy (eV) and momentum (eV/c) of (*a*) the electron ? (*b*) the proton ?

Sol. (a) In the photoelectric effect, the photon (γ) gives all its energy and momentum to the bound system :

$$E_\gamma\,(\text{eV}) = \frac{12\,400 \text{ eV·Å}}{\lambda\,(\text{Å})}$$

$$= \frac{12\,400 \text{ eV·Å}}{2.00 \text{ Å}} = 6200 \text{ eV}$$

$$p_\gamma = \frac{E_\gamma}{c} = 6200 \text{ eV/c}$$

Some of the energy, 14 eV, is used in unbinding the electron and proton. Most of the remaining energy, (6200 – 14) eV=6186 eV, will appear as kinetic energy of the released electron because it is much lighter than the proton. Thus, E_k (electron) $\approx 6.19 \times 10^3$ eV. Since this is much less than the electrons rest energy, $E_0 = 5.11 \times 10^5$ eV, we can treat the electron classically to get its momentum :

$$E_k = \frac{p^2}{2m_0} = \frac{(pc)^2}{2E_0}$$

or $$p_e c = \sqrt{2E_k E_0} = \sqrt{2\,(6.19 \times 10^3 \text{ eV})\,(5.11 \times 10^5 \text{ eV})}$$

$$p_e = 7.95 \times 10^4 \text{ eV/c}$$

(b) By conservation of momentum, $p\gamma = p_e + p_p$, or

$$p_p = (6.20\times10^3 - 7.95\times10^4)\ \text{eV}/c$$
$$= -7.33\times10^4\ \text{eV}/c$$

The proton recoils in a direction opposite to that of the electron and with a momentum of great magnitude. Its kinetic energy is

$$E_k\,(\text{proton}) = \frac{p^2}{2m_0} = \frac{(pc)^2}{2E_0}$$
$$= \frac{(7.33\times10^4\ \text{eV})^2}{2(9.38\times10^8\ \text{eV})}$$
$$= 2.9\ \text{eV}$$

As we assumed initially, the proton's kinetic energy is much smaller than the kinetic energy of the electron.

Example 2.2. It is possible for the photoelectric effect to occur when the photon is incident on a single *unbound* charged particle ?

Sol. We assume that an incident photon with energy $h\nu$ and momentum $h\nu/c$ collides with a free particle of rest mass m_0 initially at rest. Suppose that the photon is annihilated and gives all its energy and momentum to the free particle. Designating the energy and momentum of the *particle* after collision by E and p, we have

Conservation of energy : $h\nu = E - E_0$

Conservation of momentum : $\dfrac{h\nu}{c} = p = \dfrac{\sqrt{E^2 - E_0^2}}{c}$

Substituting $h\nu$ from the first equation into the second, and rewriting, we get

$$E - E_0 = \sqrt{E - E_0}\,\sqrt{E + E}$$

or $$\sqrt{E - E_0} = \sqrt{E + E_0}$$

which is true only if $E_0 = 0$. Since any material particle has a nonzero rest mass, $E_0 \neq 0$, the photoelectric effect cannot take place with a free particle. In an actual photoelectric effect, one with bound electrons, the electron's mass is much smaller than the mass of the object to which it is bound. Then nearly all the photon's energy is transferred to the electron, and only a small fraction to the object to which the electron is bound. On the other hand, only a fraction of the photon's momentum is transferred to the bound electron.

2.5. Compton Effect

In the photoelectric effect, a photon gives all its energy to a bound electron-metal system; it is also possible for only a part of a photon's energy to be given to a charged particle. This kind of interaction between electromagnetic waves and a material substance is a *scattering* of the waves by the charged particles of the substance. The quantum theory of the scattering of electromagnetic waves is known as the *Compton effect.* We shall first review briefly the classical theory of the scattering of electromagnetic waves by charged particles.

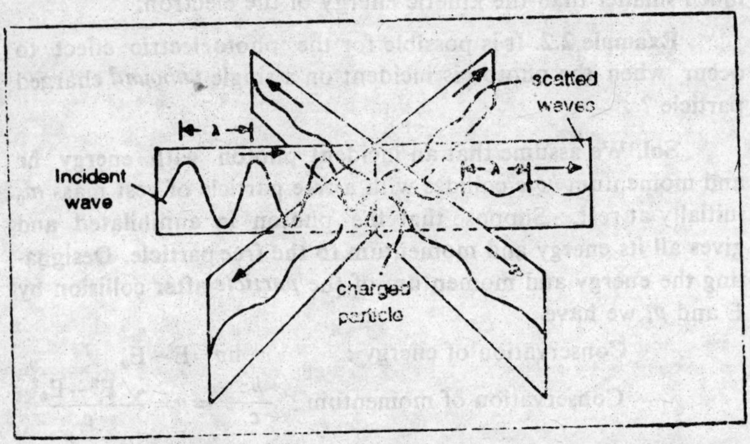

Fig. 2.8. Classical scattering of electromagnetic radiation by a charged particle.

When a monochromatic electromagnetic wave impinges upon a charged particle whose size is much smaller than the wavelength of the radiation, the charged particle will be acted on principally by the sinusoidally varying electric field of the wave. Under the influence of this changing electric force, the particle will oscillaie in simple harmonic motion at the frequency of the incident radiation (see Figure 2.8). Since the charge is accelerated continuously, it will produce electromagnetic radiation of the *same* frequency in all directions; the intensity is greatest in the plane perpendicular to the direction of motion of the oscillating charge and zero along the line of oscillation.+ Classical theory predicts, then, that the scattered

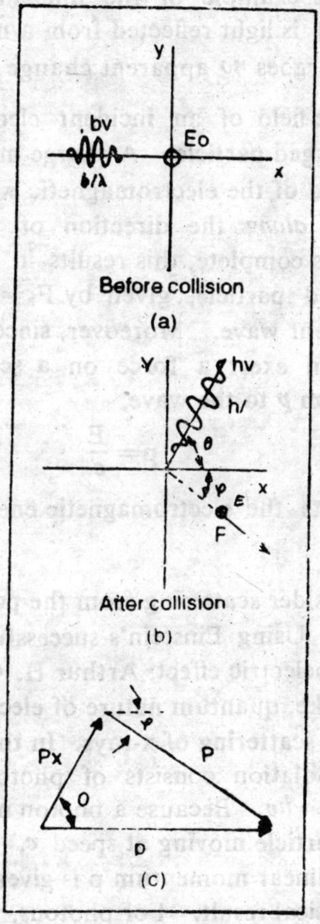

Fig. 2.9. Collision of a photon and a particle initially at rest.

radiation will have the *same* frequency as the incident radiation. The charged particle becomes the transfer agent, absorbing some of the energy from the incident beam and reradiating this energy at the same frequency (or wavelength), although scattering it in all directions. The scattering particle neither gains nor loses energy, since it reradiates at the same rate as it absorbs. The classical scattering theory agrees with experiment for wavelengths of visible light and all other longer wavelengths of radiation. A simple example of the unchanged frequency of scattered radiation is light reflected from a mirror (a collection of scatters); it undergoes *no* apparent change in frequency.

The magnetic field of an incident electromagnetic wave also affects a charged particle. A charge moving in the transverse magnetic field of the electromagnetic wave is acted on by a magnetic force *along* the direction of wave propagation. When absorption is complete, this results in a radiation force F_r on the charged particle, given by $F_r = P/c$, where P is the power of the incident wave. Moreover, since a classical eletromagnetic wave can exert a force on a scattering center, we attribute momentum *p* to the wave,

$$p = \frac{E}{c}$$

where E represents the electromagnetic energy of the incident wave.

Now we consider scattering from the point of view of the quantum theory. Using Einstein's successful photon interpretation of the photoelectric effect, Arthur H. Compton in 1922 used the particle like, quantum nature of electromagnetic radiation to explain the scattering of x-rays. In the quantum theory, electromagnetic radiation consists of photons. each with an energy given by $E = h\nu$. Because a photon may be regarded as a zero-rest-mass particle moving at speed c, the magnitude of the corresponding linear momentum p is given by E/c, in agreement with the classical result. For photons, we then have

$$p = \frac{E}{c} = \frac{h\nu}{c} = \frac{h}{\lambda} \qquad (9)$$

Each photon in a beam of monochromatic electromagnetic radiation of wavelength λ has a momentum equal to h/λ. Equation 9 shows that the momentum of a photon is precisely specified when the wavelength, the frequency, or the energy of the photon is known. The direction of p is along the direction of propagation of the wave.

The distinctive feature the quantum theory introduces is that for monochromatic waves the electromagnetic momentum occurs not in arbitrary amounts but only in integral multiples of the momentum h/λ carried by a single photon. From Equation 8 we see that a photon's momentum increases with frequency, just as its energy increases with frequency. Therefore, the momentum of a high frequency, high-energy photon, such as γ ray, will far exceed the momentum of a low-frequency, low-energy photon. such as a radio photon.

If we regard a monochromatic electromagnetic beam as consisting of a collection of particle like photons, each with a precisely defined energy and momentum, the scattering of electromagnetic radiation becomes a problem involving the collision of a photon with a charged particle. Then the problem is solved merely by applying the laws of energy and momentum conservation. Figures 2.9 a and b shows the incident photon and particle (at rest) before the collision and the moving particle and scattered photon after collision. In applying, the conservation laws, we need to be concerned not with the details of the interaction between the photon and particle during the collision, but with only the total energy and momentum going into and coming out of the collision.

Unlike the classical scattering of electromagnetic waves, in which the particle after collision is assumed to gain essentially no energy, the quantum treatment calls for it to gain some; because its kinetic energy may be large, we shall treat it relativistically.

We take the particle, of rest mass m_0 and rest energy $E_0=m_0c^2$, to be *free* and initially at rest. Then, applying energy conservation to the collision of Figure 2.9 yields

$$h_x+E_0=h\nu'+E \qquad \ldots(9)$$

Where E is the energy of the recoiling *particle* after collision, $h\nu$ is the energy of the incident photons, and $h\nu'$ is the energy of the scattered photons. Since the final energy (rest energy plus kinetic energy) of the recoil particle, $E=\gamma E_0$, must exceed its initial energy E_0, we immediately see from Equation 9 that $h\nu' < h\nu$. Consequently, the scattered photon has *less* energy, a lower frequency, and a longer wavelength than the incident photon. This disagrees with the classical prediction of no frequency change on scattering. Do not think that the scattered photon is merely the incident photon moving in a different direction. Instead, the incident photon is annihilated, and the scattered photon is created.

Momentum conservation is implied by the vector triangle of Figure 2.9c, where $p=\gamma m_0 v$ is the relativistic momentum of the recoiling particle. The magnitude of the momentum of the incident photon is $p\lambda=h\nu/c=h/\lambda$ and of the scattered photon is $p\lambda_{,}=h\nu'/c=h/\lambda'$. The scattering angle θ is the angle between the directions of $p\lambda$ and $p\lambda_{.}$, the directions of the incident and scattered photons.

We want to solve for the change in wavelength $\lambda'-\lambda=\Delta\lambda$ in terms of θ. Applying the law of cosines to the triangle in Figure 4–12c, we have

$$p\lambda^2+p\lambda_{.}^{2}-2p\lambda p\lambda_{.}\cos\theta=p^2 \qquad -(10)$$

Multiplying both sides of this equation by c^2 and recalling that for a photon $pc=h\nu$, we have

$$h^2\nu^2+h^2\nu'^2-2h^2\nu\nu'\cos\theta=p^2c^2 \qquad -(11)$$

We can arrive at a similar expression from the energy-conservation relation, Equation 9. We place $h\nu$ and $h\nu'$ on one side of Equation 4-11 and E and E_0 on the other then squaring the equation, we get

$$h^2\nu^2+h^2\nu'^2-2h^2\nu\nu'=E^2+E_0^2-2EF_0$$
$$=2E_0^2+p^2e^2-2EE_0 \qquad -(12)$$

where we have replaced E^2 with $E_0^2+p^2c^2$. Subtracting Equation 11 from Equation 12, we have

$$-2h^2\nu\nu'(1-\cos\theta)=2E_0^2-2EE_0$$
$$h^2\nu\nu'(1-\cos\theta)=E_0(E-E)=m_0c^2(h\nu-h\nu')$$

$$\frac{h}{m_0c}(1-\cos\theta)=c\frac{(\nu-\nu')}{\nu\nu'}=\frac{c}{\nu'}-\frac{c}{\nu}=\lambda'-\lambda$$

The increase in wavelength $\Delta\lambda$ then, is

$$\Delta\lambda=\lambda'-\lambda=\frac{h}{m_0c}(1-\cos\theta) \qquad -(13)$$

This forms the basic equation of the Compton effect. It gives the increase $\Delta\lambda$ in the wavelength of the scattered photon over the value of the wavelength of the incident photon. We see that $\Delta\lambda$ depends only on the rest mass m_0 of the recoiling particle, and on Planck's constant h, the speed c of light, and the angle θ of scattering. It is perhaps surprising to find that $\Delta\lambda$ is *independent* of the incident photon's wavelength λ. The quantity h/m_0c, appearing on the right-hand side of Equation 13 and having the dimension of length, is known as the *Compton wavelength*. Although the scattering angle θ determines the wavelength increase $\Delta\lambda$ unambiguously, we cannot predict in advance the angle at which any one photon will emerge.

If the recoiling particle is a free electron within the scattering material, then $m_0=9.11\times10^{-31}$ kg and $h/m_0c=0.024\ 26$ Å. When a photon emerges at, for example, $\theta=90°$ with respect to the incident-photon direction, the wavelength change, by Equation (13), is 0.024 Å. When a scattered photon emerges at $\theta=180°$, travelling, in other words, in the backward direction, with the recoil electron travelling in the forward direction, the collision can be characterized as a head-on collision. Then the wavelength change is a maximum and equal to 0.049 Å. In such a collision, the electron's kinetic energy is also a maximum.

For the 90° scattering, by a free electron, of incident radiation in the visible region, say 4000 Å, the *fractional* increase in wavelength $\Delta\lambda/\lambda$ is only 0.006 per cent. Such a shift in wavelength is completely masked in visible light since electrons in an ordinary scattering material are not at rest but in thermal motion. At room temperature this results in a spread $\Delta\lambda/\lambda\approx0.3$ per cent arising from thermal motion of electrons. An observable shift of, say, $\Delta\lambda/\lambda=2$ per cent can be obtained by using incident radiation of wavelength $\lambda=1$ Å; then $\Delta\lambda=$

0.024 Å. Thus, there is an easily observed shift for x-ray photons and photons of shorter wavelength. For photons of longer wavelength, the fractional change in wavelength is very small, and the scattered radiation has nearly the same wavelength and frequency as the incident radiation. Classically, the wavelengths of the incident and scattered radiations are equal; hence Compton scattering agrees with classical scattering in the region of $\Delta\lambda/\lambda < 1$. This is another example of the correspondence principle; by Equation (13)

$$\lim_{\substack{m_0 \to \infty \\ \text{or } h \to 0}} \frac{\Delta\lambda}{\lambda} = \lim_{\substack{m_0 \to \infty \\ \text{or } h \to 0}} \frac{h}{m_0 c \lambda} = 0$$

That the scattering of x-rays agrees with the photon model instead of the classical model (whieh predicts no change in wavelength) was shown first by A.H. Compton in 1922. Figure 2.10 gives schematically the experimental arrangement for monochromatic x-rays incident on a target of carbon, a substance having many free electrons. For any fixed angle θ, the detector can measure the scattered radiation's intensity as a function of wavelength. (Compare Figure 2.10 with Figure 2.8, where $\lambda = \lambda'$ and $\Delta\lambda = 0$). Figure 2.11 shows the intensity of the scattered radiation versus the scattered wavelength for several fixed angles θ.

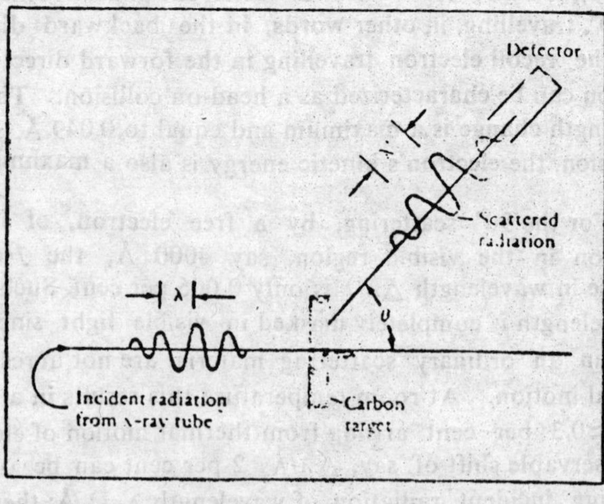

Fig. 2.10. Schematic of experimental arrangement for the Compton effect.

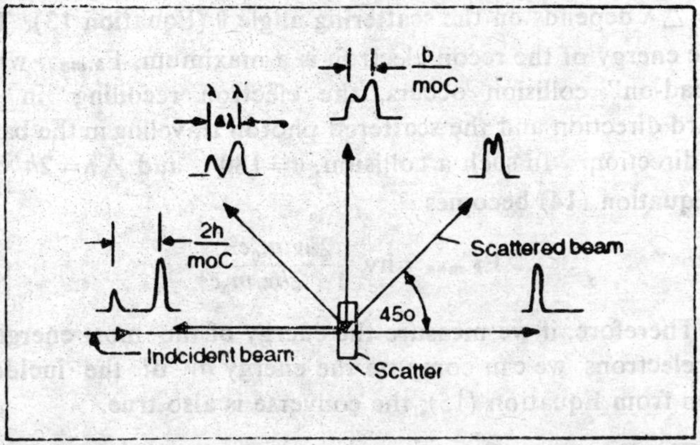

Fig. 2.11. Intensity of scattered radiation versus wavelength of scattered radiation for different angles E.

For any particular scattering angle θ, *two* predominant wavelength are present in the scattered radiation : one of the same wavelengths λ as the incident beam, the *unmodified wave*, and a second, of longer wavelength λ', the *modified wave*, given by the Compton equation (13). The unmodified wavelength results from the coherent scattering of the incident radiation by the Inner electrons of atoms; these electrons are so tightly bound to the atoms that a photon cannot strike one of them without simultaneously moving the entire atom. The mass m_0 of one of these tightly bound electrons is, then, *effectively* the mass M_0 of the atom. Therefore, in a Compton collision between a photon and a tightly bound electron, the wavelength change $\Delta\lambda$ is $(h/M_0 c)(1-\cos\theta)\approx 0$, because M_0 is always thousands of times greater than m_0.

The Compton effect provides a simple way of determining the energy of a photon. From Equation (9), we have

$$E_k = E - E_0 = h\nu - h\nu'$$

Because $\nu = c/\lambda$ and $\nu' = c/\lambda' = c/(\lambda + \Delta\lambda)$, we can write this as

$$E_k = h\nu \frac{\Delta\lambda}{\lambda + \Delta\lambda} = h\nu \frac{\Delta\lambda/\lambda}{1 + (\Delta\lambda/\lambda)} \qquad (14)$$

where $\Delta\lambda$ depends on the scattering angle θ (Equation 13). The kinetic energy of the recoil electron is a maximum, $E_{k.max}$, when a "head-on" collision occurs, the electron recoiling in the forward direction and the scattered photon traveling in the backward direction. In such a collision, $\theta=180°$, and $\Delta\lambda=2h/m_0c$; and Equation (14) becomes

$$E_{k\cdot max}=h\nu\frac{2h\nu/m_0c^2}{1+2h\nu/m_0c^2} \tag{15}$$

Therefore, if we measure the energy of the most energetic recoil electrons we can compute the energy $h\nu$ of the incident photon from Equation (15); the converse is also true.

The Compton effect shows clearly the particle like aspects of electromagnetic radiation - not only a precise energy $h\nu$ but also a precise momentum h/λ can be assigned to a photon. Along any direction, the total momentum of a monochromatic electromagnetic beam can then assume *not* any value but only an exactly integral multiple of h. The linear momentum as well as the energy of electromagnetic radiation is quantized.

Example 2.3. A 2.00-Å photon interacts with the bound electron in a hydrogen atom (binding energy, 13.6 eV). A Compton collision occurs and the electron moves forward in the same direction as the incident photon. (*a*) What is the energy of the electron ? (*b*) What is the energy of the scattered photon ?

Sol. (a) Since the energy of the incident photon,

$$E=\frac{12\,400\text{ eV.Å}}{\lambda}=\frac{12\,400\text{ eV Å}}{2.00\text{ Å}}=6200\text{ eV}$$

is much larger than the binding energy of the electron we can ignore the binding energy. Then, by Equation (15).

$$E_k=h\nu\frac{(2h\nu)/m_0c^2}{1+2h\nu/m_0c^2}$$

$$=(6.2\times10^3)\frac{(2\times6.2\times10^3)/(5.1\times10^5)}{1+[(2\times6.2\times10^3)/(5.1\times10^5)]}\text{eV}$$

$$=147\text{ eV}$$

(b) Since the electron moves forward. the scattered photon must move backward. Then $\theta=180°$, and $\Delta\lambda=2h/m_0c=0.049$ Å. Therefore,

$$\lambda'=\lambda+\Delta\lambda=2.0+0.049\approx2.049 \text{ Å}$$

$$E'_0=\frac{12\ 400}{2.049\ \text{Å}}=6052 \text{ eV}$$

These results are consistent with energy conservation; the total kinetic energy after collision is 147 eV (electron) plus 6052 eV (scattered photon)=6200 eV, which was the energy of the incident photon.

Example 2.4. Classical electromagnetic theory predicts that the radiation pressure P_r of a completely absorbed electromagnetic beam of intensity I will be $P_r=I/c$ for incidence along the normal to the absorbing surface.† Derive this relation from the quantum theory by assuming that the radiation consists of photons, each with momentum $h\nu/c$.

Sol. We designate the flux of the photon beam to be N photons per unit time per unit area perpendicular to the beam direction. Then, with an energy $h\nu$ carried by each photon in the beam, the total energy per unit time per unit transverse area, or the intensity I, is given by

$$I=N(h\nu)$$

If the beam strikes a surface that absorbs it completely, each photon transmits a momentum $h\nu/c$ to the surface. The total momentum transmitted per unit time per unit area is then $N(h\nu/c)$; but the total momentum transferred per unit time is just the radiation force on the absorbing surface, and this force per unit area is the radiation pressure P_r ;

$$P_r=\frac{Nh\nu}{c}$$

Eliminating $Nh\nu$ from the two equations, we have

$$P_r=\frac{I}{c}$$

for complete absorption. If the beam strikes a totally reflecting surface along the normal to this surface, the momentum is transferred *twice* to the surface, once when an incident photon is annihilated on striking the surface and again when a second photon of the same frequency is created and travels outward from the surface. Consequently, the radiation pressure is given by

$$P_r = \frac{2I}{c}$$

for complete reflection.

Emission of photons is the reverse of photon absorption. When a source emits photons, momentum conservation requires that the emitter recoil in the opposite direction. An example of this is the "photon rocket," which consists simply of a unidirectional source of electromagnetic radiation, like the laser. The rocket gains momentum in the forward direction because photons are emitted to the rear. Notice that a photon carries more momentum per total energy than a particle of nonzero rest energy does; that is, from $p=(E^2-E_0^2)^{1/2}/c$, we see that for a given energy E the momentum p is a maximum for $E_0=0$.

As a zero-rest-mass particle, a photon always is observed to travel at the single speed c in any reference frame. Its energy hν and momentum $h\nu/c$, on the other hand, depend on the observer's reference frame. Both quantities are proportional to the photon's frequency, and the frequency depends in turn on the reference frame by the (relativistic) Doppler effect. Thus, an observer travelling in the direction (relative to the source) in which a photon is moving finds the photon's frequency, energy, and momentum all to have decreased relative to their values measured in a reference frame at rest with respect to the source of photons.

2.6 Radiant or Electromagnetic Energy

Radiant energy, also called *electromagnetic energy*, travels at 3.00×10^8 ms^{-1} in a vacuum, for example. This is about 186,000 mis^{-1}, whioh by camparison makes an interplanetary

rocket took like a turtle.) Such energy exhibits what is known as wave motion.

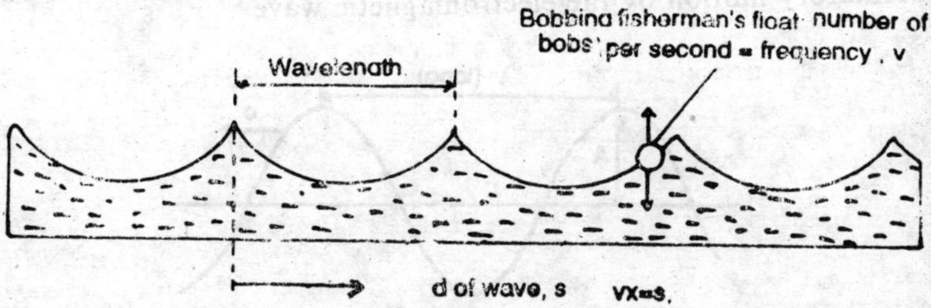

Fig. 2.12. A water wave. A fisherman's float will bob up and down as the water wave moves past. The distance between successive crests (or troughs) of the wave is called the *wavelength*. Since one wavelength passes by for each bob of the float, the product of the number of bobs per second (the frequency times the wavelength equals the distance traveled by the wave in one second, the speed of the wave.

What is wave motion? You are probably familiar with water waves, one of which is shown schematically in Fig. 2.12. As the wave moves from left to right, any floating object, such as the fisherman's float shown, bobs up and down as successive crests and troughs of the water wave pass it. When we speak of the wave, we mean the repetitive disturbance which travels across the surface of the water and causes the float to bob. The *frequency* of the wave is represented by the Greek letter nu, ν, and is the number of bobs made by the float per second. The *wavelength* is represented by lambda, λ, and is the distance between successive corresponding points, such as crests. The product of these is the *speed s* at which the wave move across the water, that is,

$$\nu\lambda = s$$

Similar of the water wave is the *electromagnetic wave*, by which radiant energy is transmitted. An electromagnetic wave is a combination of an electric and a magnetic field, both of which are oscillating and traveling through space. What does that mean? It means that a tiny, electrically charged particle will "bob" back and forth as the electromagnetic wave goes by, in much the same way as the fisherman's float bobs in response to a water wave. Also, a tiny magnet will be set into a similar oscillatory motion by the electromagnetic wave.

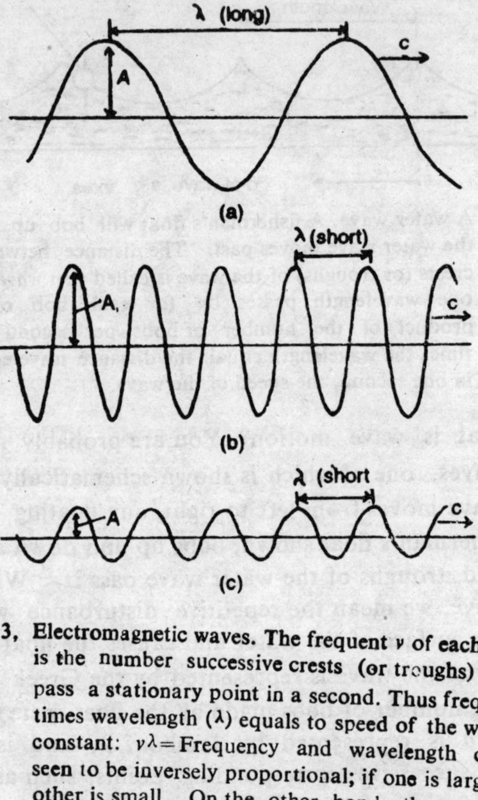

Fig. 2.13, Electromagnetic waves. The frequent ν of each wave is the number successive crests (or troughs) which pass a stationary point in a second. Thus frequency times wavelength (λ) equals to speed of the wave (a constant: $\nu\lambda=$ Frequency and wavelength can be seen to be inversely proportional; if one is large, the other is small. On the other hand, the amplitude *A* of the wave can vary indepently, as can be seen by comparing waves *b* and *c*.

Figure 2.13 schematically illustrates three examples of electromagnetic waves. The shape of each wave is that of the familiar sine wave, and the wavelength, frequency, and speed are all related as they were for the water wave. Because the speed of all electromagnetic waves is a constant, at least in a vacuum, it is designated by *c*, so that we may write

$$\nu\lambda = c$$

This means that the frequency of a wave and its wavelength are inversely proportional; as one goes up the other goes down.

The range of observed electromagnetic frequencies (and of corresponding wavelengths) is depicted in Fig. 2.14. The so-called visible spectrum is the relatively narrow band of wavelengths and corresponding frequencies which our eyes (and brains) are able to detect. This is shown expanded in the

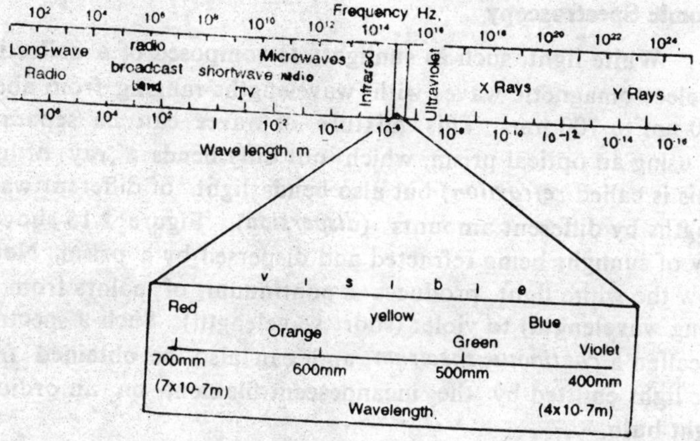

Fig. 2.14. **The electromagnetic spectrum. Energy increases from left to right in the diagram.**

diagram so that you can correlate wavelength with color. Short-wavelength (high frequency) waves we *see* as violet, for example. In Fig. 2.14 frequencies are indicated in *hertz* (Hz), a unit derived from the SI base units. It is the same as *cycles per second*, or s^{-1} (read "reciprocal seconds," "inverse seconds," or just "per second"). Wavelengths are indicated in the figure in meters (m), except for the visible spectrum where they are shown as nanometers (nm). (1 nm $= 10^{-9}$ m).

Example 2.5. A certain AM radio station broadcasts on a frequency assigned to it which is 980 kilohertz (kHz). What is the wavelength of the electromagnetic wave broadcast by the station ?

Solution. $980\ kHz = 980 \times 10^3\ Hz = 980 \times 10^3\ s^{-1}$

Rearranging the relationship $\nu\lambda = c$ to solve for λ, we get

$$\lambda = \frac{c}{\nu}$$

$$= \frac{3.00 \times 10^8 m\ s^{-1}}{9.80 \times 10^2\ s^{-1}} = 306m$$

The wavelength of this radio wave is thus 308 m, or about 335 yd.

Atomic Spectroscopy

White light. such as sunlight, is composed of a collection of electromagnetic waves with wavelengths ranging from about 400 nm to 700 nm. This mixture of waves can be separated by using an optical prism, which not only bends a ray of light (this is called *refraction*) but also bends light of different wavelengths by different amounts (*dispersion*). Figure 2.15 shows a ray of sunlight being refracted and dispersed by a prism. Notice how the white light produces a continuum of colors from red (long wavelength) to violet (short wavelength). Such a spectrum is called a *continous spectrnm* and can also be obtained from the light emitted by the incandescent filament on an ordinary light bulb.

When substances are heated to a very high temperature by a flame or an electric arc to spark, light is given off, and the color depends upon the substance being investigated. Sodium nd sodium-containing compounds give off bright yellow light, while potassium and potassium-containing compounds produce weaker violet color. Such colors can be investigated spectroscopically in order to determine what wavelengths of light are emitted. When this is done with hydrogen gas the results are, like those shown in Fig. 2.16. Instead of a continuous spectrum, a *line spectrum* is produced. Each line on the screen is, produced by light of a single discrete wavelength and frequency and the whole series of line is called the *Balmer series*, after J.J. Balmer, who studied it in 1885. Balmer discovered a surprisingly simple relationship among the different wavelengths which produce the lines in the series. One form of this relationship is

$$\frac{1}{\lambda}=R\left(\frac{2}{2^1}-\frac{1}{n^2}\right)$$

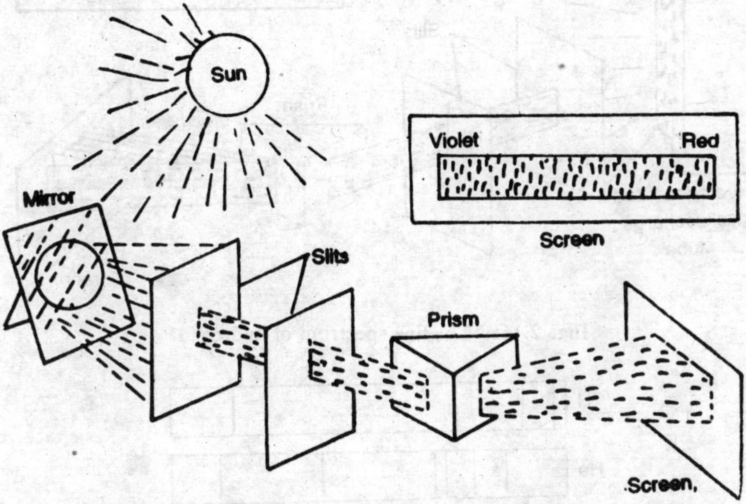

Fig. 2.15. The spectrum of sunlight (Fraunhofer lines not shown). The colors on the screen form a continuous band from violet at one end to red at the other.

Here λ is the wavelength in nanometers, R is a constant now called the *Rydberg constant*, and *n* is an integer, 3 or larger. The valve of *R* is 1.10×19^{-2} nm^{-}. Balmer found that this formula could be used to predict the wavelength of each of the lines within experimental error just by substituting integers for n. In other words each value of *n* (3 or larger) substituted in the Balmer formula yields the correct wavelength of one of the spectral lines.

Hydrogen produces other series of spectral line, such as the *Lyman series* in the ultraviolet region of the spectrum, and the *Paschen series* in the infrared. Each series can be described by an equation which generates the wavelengths of all the lines in the series, as Balmer's equation did for the Balmer series. The different equations can be combined into a single relation, sometimes called the *Rydberg* equation

$$\frac{1}{\lambda}=R\left(\frac{1}{n_1^2}-\frac{1}{n_2^5}\right) \quad ...(1)$$

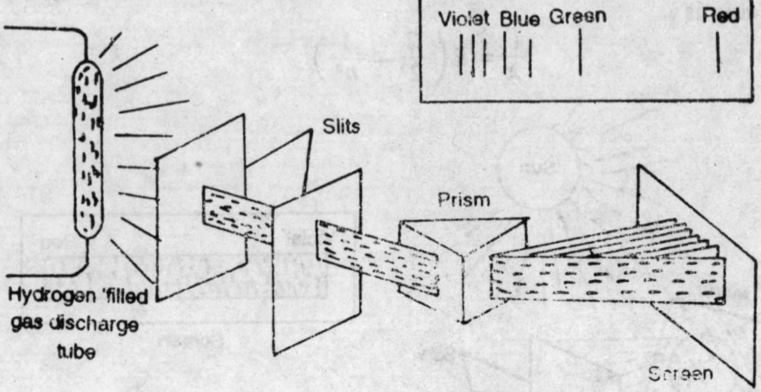

Fig. 2.16. The line spectrum of hydrogen.

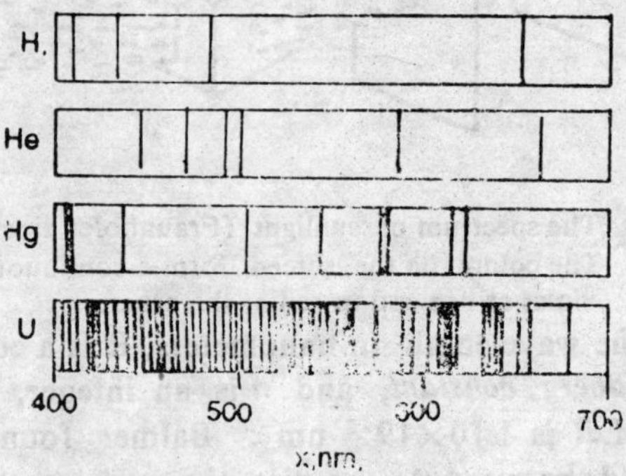

Fig. 2.17. Line spectra of some elements.

This equation generates the wavelengths of all the lines in each series in the following way :

Lyman series : $n_1=1$ $n_2=2, 3, 4, 5..., \infty$
Balmer series : $n_1=2$ $n_2=3, 4, 5, 6.. , \infty$
Paschen series : $n_1=3$ $n_2=4, 5, 6, 7 , \infty$

One more very important point should be noted : other elements besides hydrogen give line spectra, too. These spectra are more complex, consisting of more lines and of lines whose wavelengths are not related by anything as simple as the Rydberg equation. But the mere fact that elements in general exhibit line spectra proves to be very significant. The line spectra of several elements are shown in Fig, 2.17.

2.7. Hydrogen Spectrum

Since the planetary model of the atom is fundamentally defective, a correct model is needed – one that can account in detail for the observed spectra and the stability. Before we discuss the model that adequately describes the structure of hydrogen atoms, the natural starting point for any theoretical description of atomic structure, let us set down what is known about the hydrogen spectrum. To observe the spectrum of isolated hydrogen atoms, one must use gaseous atomic hydrogen, because in it the atoms are so far apart that each one behaves as an isolated system. Molecular hydrogen, H_2, and solid hydrogen radiate spectra that reflect some aspects of hydrogen atoms bound together). The visible spectrum emitted by hydrogen can be studied with a prism spectrometer: one is shown sthematically in Figure 2.18. (A diffraction grating can be used instead of a prism for dispersing the radiation.) The hydrogen gas is excited by an electrical discharge or by extreme heating and emits radiation. The dispersed radiation, separated into its various frequency components, falls on a screen or photographic plate, making it possible to measure the frequency and intensities of the emission spectrum.

Any instrument that disperses and measures the various wavelengths of a beam of electromagnetic radiation is called a spectometer. An instrument that disperses the light and photographs the spectrum is known as a spectrograph; one that makes the spectrum visible directly to the eye is called a spectroscope. Spectrometer are designed for studying each of the several regions of the electromagnetic spectrum, such as radiofrequenc,y x-rays γ rays. The branch of physics that deals with the electromagnetic radiation emitted or absorbed by substances is called spectroscopy. Spectroscopy is a very powerful method of inquiry into atomic, molecular, and nuclear, structure; it is characterized by very great precision (frequencies or wavelengths are easily measurable to 1 part in 10 million) and very high sensitivity (emission or absorption from samples as small as fractions of a microgram can be observed).

The spectrum from atomic hydrogen given by a prism or diffraction-grating spectrometer consists of numerous sharp,

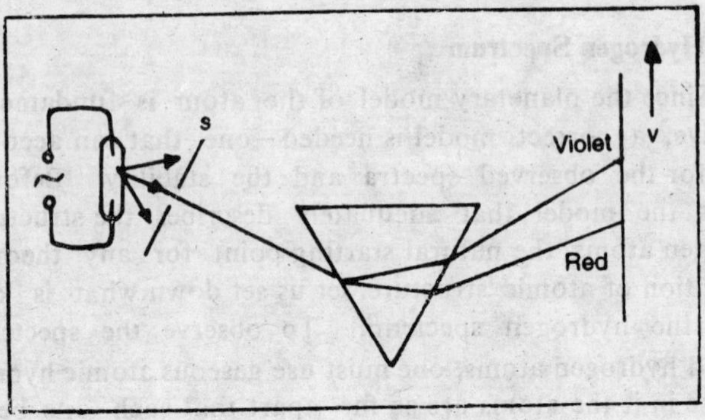

Fig. 2.18. Schematic diagram of a prism spectrometer.

discrete, bright lines on a black background; the lines are images of the slit, shown by S in Figure 2.18. The spectra of all chemical elements in monatomic gaseous from are composed of such bright lines, each spectrum characteristic of a particular element. The spectrum is known as a line spectrum. The emission spectrum from atomic hydrogen, then, is a bright-line spectrum characteristic of hydrogen. Since each chemical element has its own characteristic line spectrum, spectroscopy is a particularly sensitive method of identifying the ele ments.

The line spectrum from atomic hydrogen in the visible region is shown in Figure 2.19. The lines are labeled H_α, H_β, and so on in the order of increasing frequency (decreasing wavelength). Ordinarily the H_α line is much more intense than the H_β, which is in turn more intense than the H_γ, and so on. The spacings between adjacent lines become smaller as the frequency increases, and the discrete lines approach a *series limit*, above which there appears a weak continuous spectrum. This group of hydrogen tines, which appears in the visible region of the electromagnetic spectrum, is known as the Balmer series; in 1885, J.J. Balmer arrived at a simple empirical

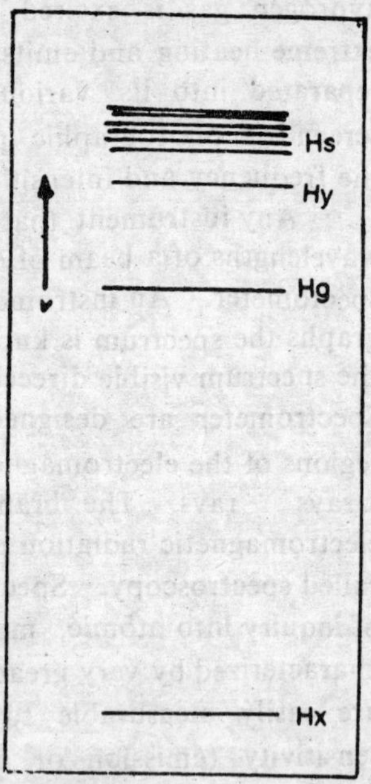

Fig. 2.19 Frequency distribution of radiation from atomic hydrogen in the visible region. The particular group of spectral lines is the Balmer series.

formula, called the Balmer formula, from which all the observed wavelengths in the group could be computed. This formula, giving the wavelength λ for all spectral lines in this series, can be written

$$\frac{1}{\lambda}=R\left(\frac{1}{2^2}-\frac{1}{n^2}\right) \quad ...(2)$$

where $R=1.0967758\times10^7\ m^{-1}\approx1.0968\times10^{-3}\ Å^{-1}$ and n is an integer having the values 3, 4, 5,...

Putting n=3 in this formula gives λ=6564.7 Å, the H_α line; similarly, putting n=4 gives λ=4862.7 Å, the H_β line, The wavelength of the series limit is given by the Balmer formula when $n=\infty$. The constant R is known as the *Rydberg constant* its experimental value is chosen by trial to give the best fit for the measured wavelengths. In atomic spectroscopy, spectral lines typically are specified by their wavelengths rather than their frequencies, because it is the wavelengths that is measured.

Besides the Balmer series in the visible region, hydrogen radiates a series of lines in the ultraviolet region and several series of lines in the infrared. Each series can be represented by a formula like the Balmer equation. In fact, one general formula can be written, from which all the spectral lines of hydrogen can be comuputed. It is known as the *Rydberg equation* :

$$\frac{1}{\lambda}=R\left(\frac{1}{n_1^2}-\frac{1}{n_u^1}\right) \quad —(3)$$

where $n_l=1$ and $n_u=2, 3, 4...$ *Lyman series*, ultraviolet region

$n_l=2$ and $n_u=3, 4, 5...$ *Balmer series*, visible region

$n_l=3$ and $n_u=4, 5, 6...$ *Paschen series*, infrared region

and so on, to further series lying in the far infrared. The value of the Rydberg constant in this equation is precisely the same as its value in equation 2 ; Equation (3) becomes (1) when

n_l is set equal to 2. The several series of hydrogen lines are named after their discoverers. Although the Rydberg formula is remarkably successful in summarizing the wavelengths radiated by atomic hydrogen, it is merely an empirical relation; in itself, it supplies no information about the structure of hydrogen. On the other hand, a truly successful theory of the hydrogen atom must be capable of predicting the spectral lines; that is, it must yield the Rydberg formula as a result.

We have been concerned with the spectrum given by hydrogen when it is excited by an electrical discharge or by extreme heating. Atomic hydrogen at room temperature does not, by itself, emit electromagnetic radiation. but at room temperature it can selectively absorb electromagnetic radiation, giving an *absorption spectrum*. The absorption spectrum of atomic hydrogen is observed when a beam of white light (all frequencies present) is passed through atomic hydrogen gas and the spectrum of the transmitted light is examined in a spectrometer. What is found is a series of dark lines superimposed on the spectrum of white light; this is known as a *dark-line spectrum*. The gas is transparent to waves of all frequencies except those corresponding to the dark lines, for which it is opaque; that is, the atoms absorb only waves of certain discrete, sharp frequencies from the continuum of waves passing through the gas. The absorbed energy is very quickly radiated by the excited atoms, but in *all directions*, not just in the incident direction. Therefore, these lines will appear dark when one observes the spectrum of the light beam emerging from the gas. The dark lines in the absorption spectrum of hydrogen and the bright lines in the emission spectrum occur at precisely the same frequencies, as shown schematically in Fig. 2,20, where the intensity is plotted as a function of frequency. Hydrogen is a radiator of electromagnetic radiation only at the specific frequencies or wavelengths given by the Rydberg formula; it is an absorber of radiation only at the same frequencies.

What holds for the emission and absorption spectra of atomic hydrogen holds equally well for the line spectra of all elements. A characterstic set of frequencies is emitted when the atoms radiate energy ; the same set of frequencies is absorbed by the atoms when a continuous frequency band of electromagnetic radiation is sent through the gas.

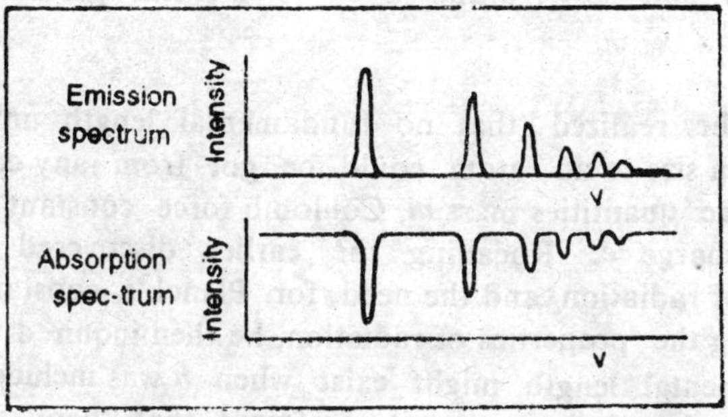

Fig. 2.20. Variation of intensity with frequency in emission and absorption spectra of hydrogen.

Bohr Theory of Hydrogen

The planetary model of the hydrogen atom cannot account for known experimental properties of hydrogen—an atomic size of the order of angstroms, the stability of hydrogen, and the discrete line spectra emitted and absorbed by a hydrogen gas. The model fails because it does not include quantum effects—the particle aspects of radiation and the wave aspects of material particles.

The first quantum theory of the hydrogen atom was proposed in 1913 by Niels Bohr, who was then working under Rutherford at the University of Manchester. Einstein has earlier established the photon nature of electromagnetic radiation, but the wave aspects of material particles were not reconized until de Broglie's work in 1924. Nevertheless, Bohr was able to make the first long step toward a thoroughgoing wave mechanical treatment of atomic structure. Although the Bohr model has limited applicability, it introduces enough quantum features to describe fairly accurately the atomic spectrum of hydrogen. Bohr's model is transitional between classical mechanics and the more complete wave-mechanical model developed during the 1920s.

Following the strictly classical model of the hydrogen atom, Bohr assumed that the more massive proton remained at rest and the electron moved about it in a circular orbit because of the electrostatic force between two charges, Thus, Equation. $mv^2 = ke^2/r$, and $E = -ke^2/2r$, still pertain in his model.

Bohr realized that no fundamental length unit comparable in size to an atom could be got from any combination of the quantities mass m, Coulomb force constant k, and electric charge e. Rocalling the earlier discovered photon nature of radiation and the need for Planck's constant h in describing the properties of radiation, he then inquired whether a fundamental length might exist when h was included. By dimensional-analysis arguments, he found that there was indeed a combination of m, k. e, and h having the unit length, namely. h^2/ke^2m and that its magnitude was of the order of atomic dimensions.

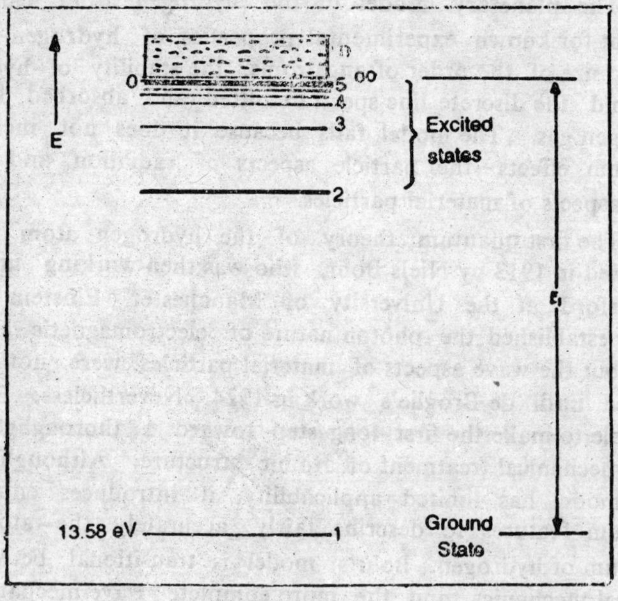

Fig. 2.21. Energy level diagram of a hydrogen atom.

To account for stable states of an atom, Bohr boldly assumed that despite the predictions of classical electromagnetism, the atom could exist in certain discrete bound states with energies E_1, E_2..... without radiating (Fig. 2.21). For unbound states ($E>0$), all energies would be allowed. He further assumed that radiation occurred when the atom made a transition between stationary energy states. He incorporated Planck's constant in the model by applying total energy conservation ; this implies that the energy lost (or gained) when the atom goes from one energy state to another is converted, through the creation (or annihilation) of a single photon, into radiation energy hv. For example, an atom, initially in an upper discrete energy E_u, make a transition to a lower discrete energy E_l with the creation and emission of a single photon of energy $h\nu$. By energy conservation,

$$h\nu = E_u - E_l \qquad ...(4)$$

Whereas the Bohr theory incorporates the particle nature of electromagnetic radiation by assuming that a single photon is created whenever the atom makes a transition to a lower energy, it gives no details of the electron's quantum jump or of the photo's creation. The situation is like that in photon-electron interactions (photoelectric effect, Compton effect, and so on) in that we did not concern ourselves with the details of the interactions but merely applied the conservation laws to the states before and after the interaction.

Each of the premitted, or quantized, energies of Fig. 2.21 corresponds to a stationary state in which the atom con exist without radiating. The lowest stationary state ($n=1$) is called the ground state. Stationary states above the ground state ($n=2, 3, 4...$) are called excited states ; an atom in an excited state tends to make a transition to some lower stationary state. It can be imagined that in a downward transition, the electron jumps suddenly from one orbit to a smaller orbit. When an atom is in some excited state and has an energy E_n, the amount by which this energy exceeds the energy of the ground state is called the excitation energy. The term binding energy denotes the energy that must be added to an atom in any state to free the bound parties and thereby make $E_n=0$. The binding energy of the atom in the ground state in the energy necessary to ionize the atom ; it will be designated E_1. Obviously, $E_1=-E_1$.

Comparing Equation (4) with experimental Lydberg equation for hydrogen Equation (5)

$$\frac{1}{\lambda}=\frac{\nu}{c}=R\left(\frac{1}{n_l^2}-\frac{1}{n_u}\right) \quad ...(5)$$

Bohr obtained an expression for the allowed bound energy states

$$E_n=\frac{Rhc}{n^2} \quad ...(6)$$

with n=1, 2,...... With n=1 in Equation 1, we have

$$E_1= \quad hcR=-E_1 \quad ...(7)$$

Therefore, Equation 6 can be written

$$E_n=-\frac{E_1}{n^2} \quad ...(8)$$

Thus, the only possible energies of the bound electron-proton system constituting the hydrogen atom are $-E_1$, $-E_1/4$ $-E_1/9$......

The radius r_n of the circular orbits follow from Equation (8)

$$r_n=-\frac{ke^2}{2E_n}=\frac{n^2ke^2}{2E_1}=n_2r_1 \quad ...(9)$$

where the ground state orbit is given by

$$r_1=\frac{ke^2}{2E_1} \quad ...(10)$$

Bohr's final task was to find theoretical values (in terms of *k*, *e*, *b*, *c*, and *m*) for E_1, r_1, and R, and to compare these with the known experimental values. This he accomplished by comparing the classical frequency f_n of the orbiting electron with the quantum frequency ν_n of the emitted photon when the atom makes a transition from a stationary energy state

E_{n+1} to the adjacent lower energy state E_n. Using Equation (9) allows us to write the orbital frequency.

$$f_n=\left(\frac{2}{\pi^2k^2e^4m}\right)^{1/2}\frac{E_1^{3/2}}{n^3} \qquad ...(11)$$

The orbital frequency decrease rapidly with increasing quantum number n.

On the other hand, when an atom makes a transition from state $n+1$ to the n state, the frequency of the emitted photon is given follows,

$$\nu=\frac{E_{n+1}-E_n}{h}=\frac{E_1}{h}\left|\frac{1}{n^2}-\frac{1}{(n+1)^2}\right| \qquad ...(12)$$

The orbital frequencies, given classically, and the radiation frequencies, given by quantum considerations, are obviously not the same for low quantum numbers, since f_n and ν are different functions of n. By the Bohr correspondence principle, however, the results of quantum physics must agree with classical physics for large quantum numbers ; that is for large n, the orbits become so large that the atom can be regarded as a macroscopic system for which classical physics yields a correct description. Requiring. then, that

$$f_n=\nu \quad \text{for } n\rightarrow\infty$$

we have, from Equations 6-22 and 6-23,

$$\left(\frac{2}{\pi^2k^2e^4m}\right)^{1/2}\frac{E_1^{3/2}}{n^3}=\frac{E_1}{h}\left|\frac{2n+1}{n^2(n+1)^2}\right|\approx\frac{E_1(2n)}{hn^4}$$

For large n, both frequencies vary with $1/n^3$, and solving the equation for E_1 gives

$$E_1=\frac{k^2e^4m}{2h^2} \qquad ...(13)$$

where $h=h/2\pi$.

The numerical value of the ionization energy for the hydrogen can now be computed from the known values of the atomic constants on the right side of Equation (13)

$$E_1 = \frac{(8.98755 \times 10^9 \text{N-m}^2/\text{C}^2)^2 (1,60219 \times 10^{-19}\ \text{C})^4 (9.10953 \times 10^{-31}\ \text{kg})}{2(1.04559 \times 10^{-34}\ \text{J-s})^2 (1.60219 \times 10^{-19} \text{J/eV})}$$

$$= 13.6358\ \text{eV},$$

This is in excellent agreement with the experimental value. 13.60 eV.

The Rydberg constant expressed in terms of atomic constants follows for Equation (7)

$$R = \frac{E_1}{hc} = \frac{k^2 e^4 m}{4\pi b^2 c} \qquad \ldots(14)$$

Since the known values of the physical constants in this equation; we find that $R = 1.497\ 37 \times 10^{-3}$ Å^{-1}, which closely agrees with the experimental spectroscopic value of 1.09678×10^{-3} Å^{-3} given at Equation 6.14.

The theoretical value of the electron's orbital radius r_1 in the lowest energy state, $n = 1$, follows from Equations (10) and

$$r_1 = \frac{ke^2}{2E_1} = \frac{h^2}{km\ e^2} \qquad \ldots(15)$$

Substituting the known atomic constants of Equation (15) gives the theoretical value for the smallest allowed orbital radius, the so-called radius of the first Bohr orbit, $r_1 = 0.528$ Å. The Bohr model predicts that the size of a hydrogen atom with The smallest stationary orbit is of the order of 1Å, agreeing well with experimental determinations.

The orbitai speed of the electron in the stationary states follows from Equation (9) and (15)

$$mv_n^2 = \frac{ke^2}{r_n} = \frac{ke^2 mke^2}{n^2 h^2}$$

$$v_n = \frac{1}{n}\frac{ke^2}{h} \qquad \ldots(16)$$

This equation can be written

$$v_n = \frac{v_1}{} \qquad \ldots(17)$$

where $v_1=ke/^2h$. The permitted orbital speeds are $v_1, v_1/2$, $v_1/3$...... the electron having a maximum speed v_1 in the first Bohr orbit. The ratio of this speed to the speed of light, v_1/c, is represented by the symbol α. From Equation (16), we have

$$\alpha=\frac{v_1}{c}=\frac{ke^2}{hc} \qquad ...(18)$$

Inserting the known values of the constants in the right-hand side of this equation shows that $\alpha=1/137.0360$. Thus the electron in the first Bohr orbit moves at 1/137 the speed of light. The quantity a, a dimensionless quantity that appears frequently in the theory of atomic structure, is known as the *fine-structure constant.* To treat the Bohr hydrogen atom as a non-relativistic problem is not unreasonable; but because of the very high precision of wavelength measurements in spectroscopy, relativstic effects can be observed and so must be included in a more complete theory.

The allowed values of the orbital angular momentum $L=mvr$ for circular orbits of the electron in the hydrogen atom are particularly simple. Substituting for v_n and r_n from Equations (16) and (17)

$$L_n=mv_nr_n=m\left(\frac{ke^2}{nh}\right)\left(\frac{n^2h^2}{kme^2}\right)$$

$$L_m=uh \qquad ...(19)$$

The orbital angular momentum of the electron has only those values that are integral multiples of h,. Bohr recognized the importance of the quantization of angular momentum ; and both energy and angular momentum quantization are very important in the quantum description of atomic and nuclear systems.

The observed spectral lines of hydrogen can now be interpreted by the energy-level diagram, Fig. 2.22. The vertical lines represent transitions between stationary states ; the lengths o. these lines are proportional to the respective photon energies and therefore to the frequencies. The lines of the Lyman series correspond to those photons produceed when hydrogen atoms in any of the excited states in which n_u=2, 3, 4, 5... undergo transitions to the ground state, in which n_1=1. Transitions from

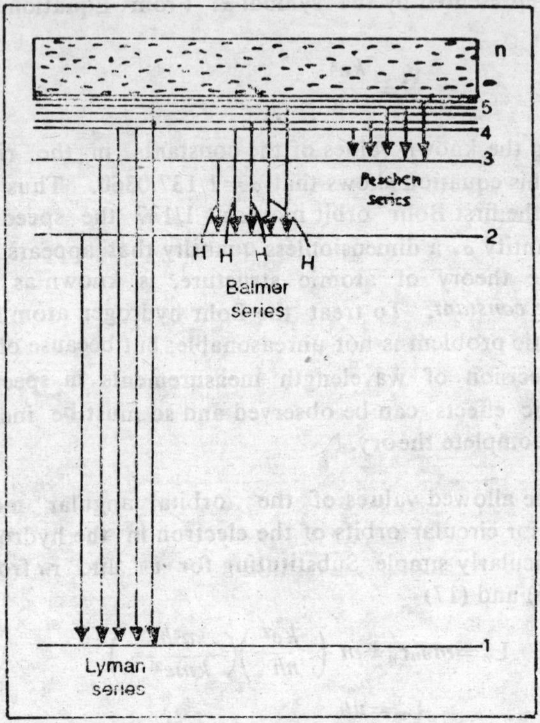

Fig. 2.22. Some possible energy transitions in atomic hydrogen.

the unbound states (E > 0) to the ground state account for the observed continuous spectrum lying beyond the series limit. We account in a similar way for the Balmer series, which is produced by transitions from the excited states in which n_u= 3, 4, 5,... to the first excited state in which n_1=2. Still further emission series involve downward transitions to n_1=3, n_1=4, and so on, these series falling progressively toward longer wavelengths.

We have examined the emission from a single hydrogen atom. The atom can exist in only *one* of its quantized energy states in any one time, and when it makes a transition from one state to a lower state, it emits a *single* photon. When the entire emission spectrum from an excited hydrogen gas, a collection of vary many hydrogen atoms, is observed in a spectroscope, we see the simultaneous emission of many photons produced by downward transitions from each of the excited states. Therefore, to observe the entire emission spectrum, we must have very many hydrogen atoms in each of the excited states making downward transitions to all lower states.

Now we also have a basis for understanding adsorption spectra. As shown in Fig. 2.20, an adsorption spectrum shows dark lines on a white background with the same wavelengths as the bright lines on the black background of the corresponding emission spectrum. When white light, consisting of photons having all possible frequencies, or energies, passes through a gas, those particular photos that have energies equal to the energy difference between stationary states can be removed from the beam. They are annihilated, thereby giving their radiant electromagnetic energy to the internal excitation energy of the atoms. The same set of quantized energy level participates in both emission and adsorption ; for this reason the frequencies of the emission and adsorption lines are identical. (Because atoms remain in an excited state for only a very short time, the Lyman series is the only one observed in adsorption).

Finite Nuclear Masses and Hydrogenic Atoms

It has been assumed thus far that the nuclear mass, compared with the electron mass, is effectively infinite and that the nuclear charge has the magnitude e. It is a simple matter to extend the Bohr theory to include (1) *finite* nuclear masses and (2) *hydrogenic* atoms, or ionized atoms in which a single electron moves about a nucleus of charge $+Ze$.

1. In an isolated atom, both the electron of mass m and the nucleus of mass M are in motion relative to atom's center of mass. which remains at rest. To take into account the motion of the nucleus, we can merely replace the electron's mass m, wherever it appears in a relation, with the so-called reduced mass, $\mu = m/(1+m/M)$. Thus, the Rydberg constant for an atom with nuclear mass M becomes, from Equation 61-4, $RM = R_{\infty}/(1+m/M,)$ where R_{∞} represents the Rydberg constant in Equation (14). For example. for hydrogen ^1H, with a single proton of nuclear mass $M_{11} = 1.67265 \times 10^{-27}$ kg, it is $R_{11} = 1.096.77 \times ^{-} Å^{-1}$, in excellent agreement with the experiment value of Rydberg's constant, 1.09678×10^{-3} $Å^{-1}$. On the other hand, for deuterium, ^2H, or heavy hydrogen, whose nucleus of mass $M_H = 3.34336 \times 10^{-27}$ kg consists of a proton and a netron bound together, the Redberg constant is $R_D = 1.09707 \times 10^{-3}$ $Å^{-1}$. Although the difference in the Rydberg constants of ^1H and ^2H is small, it leads to corresponding charges in the wavelengths of the emitted and absorbed radiation. For ^1H, for example. the Hα line is 6564.70 Å ; for ^2H it is 6562.94 Å. The diffrence in wavelength in the Hα line can be seen. Indeed, deuterium was first discovered, in 1932, when closely spaced pairs of hydrogen spectral lines were seen.

2. A hydrogenic atom has a single electron bound to a nucleus whose charge is $+Ze$. Here Z represents the atomic number, or the number of photons in the nucleus. For example, doubly ionized lithium, Li^{2+}, with $Z=3$, is a hydrogenic atom. To find the energies. frequencies, and emitted wavelengths of hodrogenic atoms, one merely replaces the quantity e^2, wherever it appears in a relation given by the Bohr theory, with Ze^2.

Successes and Failures of the Bohr Theory

The fundamental postulates of the of Bohr theory, succinctly expressed, are :

1. *A bound atomic system can exist without radiating only in certain discrete stationary states.*
2. *The stationary states are those in which the orbital angular momentum, mvr, of the atom is an integral multiple of h, Plank's c˜nstant divided by 2π.*
3. *When an atom undergoes a transition from an upper energy state E_u to a lower energy state Et, photon of energy hv is emitted, the conservation of energy requiring that $hv = E_u - E_l$: if a photon is absorbed the atom will make a transitton from the lower to the higher energy state, according to the same relation.*

These postulates are retained in their essential forms in the more complete wave-mechanical treatments of atomic and nuclear structure. The Bohr theory accounts for (1) the stability of atoms, (2) the wavelengths of the emission and absorption spectra of hydrogenic atoms. and (3) the measured ionization energies of one electron atoms.

The Bohr atomic theory has, however, certain serious shortcomings : (1) It is nonrelativistic, (2) It gives no method of calculating the intensities of the spectral lines. (3) It can't explain the spectra of atoms having more than one electron. (4) It does not explain the binding of atoms in molecules, liquids, and solids. (5) Even for hydrogen. it fails to account for the fine details of the spectrum. (High-resolution spectrographs show that each "line" predicted by the Bohr theory consists of two or more very closely spaced lines, of fine structure.) (6) It gives for orbital angular momentum—where taking into account its quantization—a rule less complicated than what physical systems actually follow.

All these defects are corrected in a relativistic wave-mechanical treatment. A fundamental reason that the Bohr thery is defective is that it overemphasizes the classical particle nature of the electron. The electron is considered to move in a well-defined circular path, the radii, speeds, and orbital frequencies being precisely defined. In a wave-mechanical treatment, the electron is a three-dimensional wave that extends and moves throughout the whole region of space surrounding the nucleus. We shall shortly consider some simple and special examples of fitting three-dimensional electron waves, by applying the Schrodinger equation, to find the quantized stationary states of hydrogen.

3

Quantum Effects : The Wave Aspects of Material Particles

3.1. De Broglie Waves

Electromagnetic radiation has two aspects, a wave aspect and a particle aspect. Experiments that show the interference and diffraction of electromagnetic radiation can be explained only if the radiation is assumed to consist of waves having frequency ν and wave length λ. The distinctively quantum effects of electromagnetic radiation, such as the photoelectric and Compton effects, can be explained only if light is assumed to consist of particle like photons, each with energy E and momentum p. The wave and particle properties of the radiation are related by :

$$E = h\nu \quad \text{.. (1)}$$

$$p = \frac{h}{\lambda} \quad \text{...(2)}$$

In these equations, the two particle like quantities appear on the left and the two wavelike quantities appear on the right. Thus, the wave-particle duality of electromagnetic radiation is implied in these relations. The fundamental constant of the quantum theory, Plank's constant h, relates the wave to the particle characteristics. We can say that electromagnetic radiation under some circumstances will behave like particles (photons of zero rest mass) and under other circumstances like waves.

Does this dual wave-particle description, found necessary for radiation, have an even greater generality, holding for *all* particles—particles with a *finite* rest mass as well as those with zero rest mass ? Louis de Broglie first posed this question in 1924. He conjectured, acknowledging the symmetry of nature, that a material particle might also show wave properties. De Broglie assumed further that the equations that give the particle characteristics of electromagnetic waves give also the wave characteristics of material particles, such as electrons. Although he did not develop quantum-mechanical equations describing general wave-particle behaviour of a material particle experiments have emphatically confirmed his hypothesis the wave character of material particles is now well established. Because the wavelength can be measured from interference or diffraction effects, we shall concentrate our attention on the second of the relations, Equation 2.

For nonrelativistic, free particles, the wavelength λ of a material particle having a momentum $p=mv$ (where m is the particle's rest mass and v its velocity) is, Equation 2.

$$\lambda=\frac{h}{p}=\frac{h}{mv} \qquad ...(3)$$

This is known as the *de Broglie relation.*

One might well ask, If an electron is to be regarded as a wave, at least under some circumstances, what is it that is waving ? That same kind of question was asked about the fundamental nature of light. Not until after the electromagnetic theory of Maxwell (1865) add the subsequent experiments of Hertz could physicists assert that the wave properties of light corresponded to oscillations of the electric and magnetic fields. Ignorance of light's electromagnetic nature, however, did not prevent physicist long before Maxwell and Hertz, for instance Young in 1801, from discovering the wavelike properties of light and interpreting interference and diffraction of this basis. Therefore, to establish whether a material particle has a wave nature, it is not necessary to know first what the wave phenomenon is. To test the de Broglie hypothesis is to determine by experimental whether material particles show interference and diffraction effects. Although the physical nature of a material particle's wave aspect is a critical consideration, we shall postpone it until after after we have discussed the experiments that confirm that a material particle has a wavelength $\lambda = h/mv$.

The first particle with finite rest mass experimentally shown to have wave properties was the electron. J.J. Thomson discovered the electron as a particle in 1897. It was found to follow well-defined paths and to have a well-defined charge-to-mass ratio (for $v \ll c$), as well as mass, momentum, and energy that could be localized in space. In short, electrons exhibit the attributes of a particle.

The wave nature of electrons was not discovered until 1927, when the electron-diffraction experiments of C. Davisson and L.H. Germer confirmed the de Broglie relation. Why were the wave characteristics of electrons discovered only many years after their particle nature had been established? The difficulties in experimentation came from the very small wavelength for ordinary particles. As. Equation 3 shows, a 1.0-kg partitle moving at 1.0 m/s has a wavelength of only 6.6×10^{-34} m $= 6.6 \times 10^{-24}$ Å. Clearly, we should not expect, in throwing baseballs through an open window, to find a discernible diffraction pattern of hits on a distant wall any more than we should expect to see one when visible light passes through such a wide "slit." The wavelength of an ordinary material particle is very small compared with the dimensions of ordinary apertures, so that interference and diffraction effects are subtle. Therefore, if an object's wavelength is to be large enough to produce observable wave effects, its mass and velocity, as we see from Equation 3, must be small.

Tne diffraction grating having the smallest distance between "lines" is a crystal. a solid in which the atoms are located in a three-dimensional geometrical array. A typical distance between adjacent atoms is of the order of 10^{-10} m, or 1 Å. The most favorable conditions for observing the diffraction of particles are those in which the particle has a wavelength of comparable size. Since $\lambda = h/mv$, to have λ large we must choose a particle with small mass; the electron is such a particle.

Let us compute the kinetic energy of an electron with a wovelength λ of 1.00 Å. An electron with electric charge *e*, accelerated from rest by an electrostatic potential difference *V*. acquired a final kinetic energy $\frac{1}{2}mv^2$ when

$$eV=\frac{1}{2}mv^2=\frac{1}{2m}\,p^2=\frac{1}{2m}\left(\frac{h}{\lambda}\right)^2$$

$$V=\frac{h^2}{2m\lambda e^2}$$

$$=\frac{(6.62\times10^{-34}\ \text{J.s})^2}{2(9.11\times10^{-31}\ \text{kg})\ (1.60\times10^{-19}\ \text{C})\ (1.00\times10^{-10}\ \text{m})^2}$$

$$=150\ \text{V}$$

An electron of 150 eV has a wavelength of 1 Å. Since this waveler gth is comparable to that of a typical x-ray photon, we can expect both electrons and x-rays to show similar diffraction effects when passing through a crystal.

3.2. Probability Interpretation of De Broglie Waves

The wave nature of electromagnetic radiation can be describrd by oscillating electric and magnetic fields in space; the wave associeted with a photon is this electromagnetic field. What about the waves associated with a material particle ? What is it that is waving when we say that an electron or any other material particle shows wave properties ? To answer this, we first relate the wave and particle descriptions of electromagnetic radiation.

When the intensity of a monochromatic beam of electromagnetic radiation directed at a screen is great, the screen appears to the eye to be uniformly illuminated; equivalently, a photographic plate placed at the screen will show a uniform darkening after it has been exposed and developed. The high intensity of the light beam means that so many photons arrive at the screen that they obscure the essentially granular and discrete nature of the electromagnetic radiation; the distinct and randomly arranged bright flashes merge into a seemingly continuous and constant illumination.

The intensity I of the illumination, the energy per unit area per unit time, is given by

$$I=\epsilon_0 E^2 c$$

where E if the magnitude of the instantaneous electric field at any point on the screen and ϵ_0 is the electric permittivity of free space. Suppose that the intensity of the beam is made extremely weak. Instead of the uniformly illuminated area observed with the high-intensity beam, distinct bright flashes appear randomly on the plate; each bright flash corresponds to the arrival of a single photon. Neither the position nor the time at which a single photon will strike the screen can be determined. Although the distribution of photons is completely random, the *average* number of photons arriving per unit area per unit time *can* be predicted; this number is the photon flux N. Since $h\nu$ is the energy per photon, the intensity of a monochromatic beam in terms of the photon flux is

$$I=(h\nu)N$$

If the positions of the individual flashes are recorded over a long period, the screen is again found to be uniformly covered.

This description of photon activity is like the kinetic theory of gases, which attributes the apparently continuous pressure of a gas to the effect of individual molecules. The molecules strike the walls of a container randomly and discretely, certaining an effect of continuous pressure by their large numbers.

The intensity of monochromatic electromagnetic radiation can be expressed by either the wave description $I=\epsilon E_0^2 c$ or the particle description $I=h\nu N$. We have in the intensity a quantity that has a precise meaning in both descriptions The intensity bridges the gap between the two disperate models.

Let us see what new meaning can be assigned to the square of the electric field strength E^2 in the photon description of light. Equating the two expressions for the intensity gives.

$$I=h\nu N=\epsilon_0 E^2 c$$

Since the probability of observing a photon on a unit area per unit time is proportional to N, the probability of observing a photon $\propto$ N $\propto$ E^2.

The probability of observing a photon at any point in space in proportional to the square of the electric field strength at that point.

By the quantum theory, the electric field is not only the quantity ihat gives the electric force per unit electric charge but also the quantity, or function, whose square gives the probability of observing a photon at any given place. Classical electromagnetic theory can yield, through computed values of E^2, the probability of observing photons, although it is incapable of yielding the strictly quantum features of electromagnetic radiation.

We can now give meaning to the wave nature of a material particle, like an electron, in the following way. We assume that the relation between the probabillty of observing a particle and the square of the amplitude of its wave is exactly analogous to the relation between the probability of observing a zero-rest-mass photon and the square of the amplitude E^2 of its wave (the electric field). The amplitude of the wave associated with a particle is represented by ψ called simply the *wave function.*

The wave function ψ is the quantity whose square ψ^2 is proportional to the probability of observing a material particle.

Thus, if ψ represents the wave function at the location x, the probability that the particle could be observed between x and x+dx is given by $\psi(x)^2\,dx$;

Probability of observing a particle in the interval dx $\propto \psi^2\,dx$.

The wave function of a particle, then, is analogous to the electric field of a photon. Just as E will generally be a function of both position and time, so will the wave function ψ.

It is not possible to specify the exact position of a *photon* at a particular time with complete certainty, but it is possible to specify the probility E^2 of observing it. Similarly it is not possible to specify the exact position of a *particle* at a particular time with complete certainty, but by ψ^2 one can specify the probability of observing it. Thus, a particle's wave function essentially gives rise to a *probability interpretation* of the position of a particle.

Tbe interpretation of waves associated with material particles through probabilities was first given in 1926 by Max Born. The branch of quantum physics that deals with finding the values of ψ is known as *wave mechanics*, or *quantum mechanics.* The two principal originators of the wave mechanics of particles were Erwin Schrodinger (1925) and Werner Heisenberg (1925), who independently formulated quantum mechanics in different but but equivalent mathematical forms.

Just as the electromagnetic theory of Maxwell is summarized in the Maxwell equations, which are the basis for computing values of E, the wave mechanics of matter is governed by the *Schrodinger equation*, which is the basis for computing values of ψ in any problem in quantum physics. Here the parallel stops, however. Whereas the electric field, which has its origin in electric charges, gives not only the probability of observing a photon but also the electric force on a unit positive electric charge, the wave function of the Schrodinger equation has a physical meaning *only* for probability interpretation – it does *not* indicate any sort of force. The wave function is not directly measureable or obserable; it does, however, give the most information one can extract concerning any system of objects; and *all* measurable quantities, such as the energy and momentum, as well as the probability of location, can be found from it.

Interference experiments have been performed with a Michelson interferometer and very weak sources of light, so weak that generally only a *single photos* was to be found between the light source and the observation screen at any one time. In the Michelson instrument, the interference pattern arises from the interference between *two* light beams. The light follows two paths going at right angles to one another from a partially silvered mirror. We might imagine that when a single photon passes through the instrument, it travels along only one of the two possible paths; yet experimental results, from data collected over a long time, show the customary interference pattern of alternating light and dark bands, implying that the single photon has travelled both routes simultaneously and has interferred with itself. The complementarity principle resolves the apparent paradox. If we speak of the photon as travelling along one of two passible routes, we are localizing it, regarding it as a particle, and precluding consideration of its wave aspects; on the other hand, if we speak of its interference pattern, we are regarding it as a wave. And to regard it both ways at once is meaningless.

A similar situation arises which waves pass through two parallel slits. When either one of the two slits is closed, the pattern is the typical single-slit diffraction pattern; a broad, central maximum flanked by weaker, secondary maxima. When both slits are open, the pattern is as shown in Figure 3.1 :

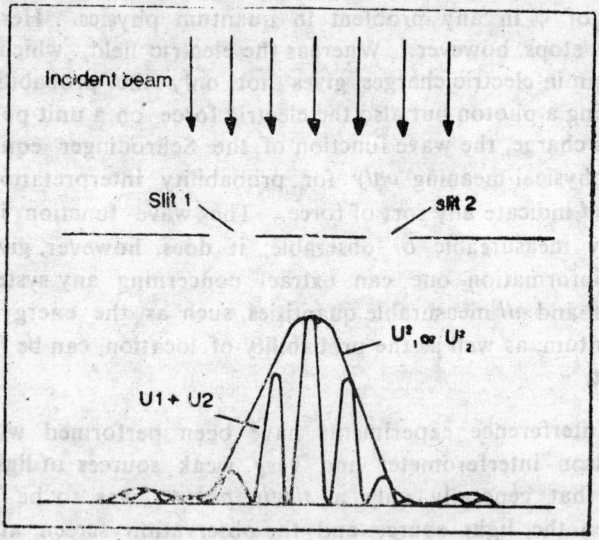

Fig. 3.1. Double slit diffraction of particles. The wave functions ψ_1 and ψ_2 give the diffraction pattern when either slit 1 or slit 2 is open, the superposed wave function ψ_1 ψ_2 gives the pattern when both slits are open. (The distance between the slits is grossyl exaggeated in the figure).

interference fine structure within a diffraction envelope. The pattern is not merely two single-slit diffraction patterns superposed; the interference between waves travelling through *both* of the slits is responsible for the rapid variations in intensity. In short, in a case in which waves can take two or more routes from a source to an observation point, we solve the problem by first superposing the wave functions (or electric fields for photons) from the two separate routes to find the resultant wave function (or electric field) and then squaring to find the probability (or intensity). That is to say, if ψ_1 and ψ_2 represent the wave function for passage through slits 1 and 2 separately, then $(\psi_1+\psi_2)^2$, not $\psi_1^2+\psi_2^2$, gives the probability of observing a particle on the screen. Thus if a single electron or proton is directed toward a pair of slits, we cannot say which of the two slits it will pass through; we must speak in the language of waves and say that in effect it passes through *both* slits.

3.3. Uncertainty Principle

The principle of complementarity shows that it is impossible to apply simultaneously the wave and particle descriptions to a material particle or to a photon. If we choose one description, we preclude the other. If we describe, for example, electromagnetic radiation in the language of particles and locate a photon at any instant with complete precision, then the *uncertainties* in position and time are both zero, $\Delta x=\Delta t=0$; on the other hand, the uncertainties in the photon's wave attributes, wavelength λ and frequency v, are then infinitely great, $\Delta\lambda=\infty$ and $\Delta v=\infty$.

Now consider a less extreme situation, one in which we are content to locate a photon in position and time, not with complete certainty, but with finite uncertainties Δx and Δt.

Earlier analysis showed that if the frequency v of a wave is measured for only a *finite* time interval Δt, then the frequency is uncertain by an amount Δv:

$$\Delta v < t \geqslant 1 \qquad -(6)$$

Similarly, if we wavelength λ is measured over a finite distance Δx along the direction of wave propagation, the wavelength is uncertain by an amount $\Delta\lambda$;

$$\Delta x \Delta \lambda \geqslant \lambda^2 \qquad \text{—(7)}$$

Note that these relations were derived on a strictly classical basis. The important implication of these relations for quantum physics is that we must associate two particlelike properties (energy E and momentum p) with two wavelike properties (frequency v and wavelength λ) according to Equations (1) and (2) :

$$E = hv \quad \text{and} \quad \frac{h}{\lambda}$$

Taking the differential of the first equation gives the uncertainty ΔE in energy in terms of the uncertainty Δv in frequency,

$$\Delta E = h \Delta v \qquad \text{—(8)}$$

Eliminating Δv from Equations (1) and (2), we have

$$\Delta E \Delta t \geqslant h \qquad \text{—(9)}$$

Thus, the product of the uncertainties in energy and in time is at least as large as Planck's constant h. The meaning, in words of Equation 9 is as follows. If an object—a photon, an electron, or even a system of particles—is known to exist in a state of energy E over a limited period Δt, then this energy is uncertain by at least an amount $h/\Delta t$; therefore, the energy of an object or system can be given with infinite precision ($\Delta E = 0$) only if the object or system exists for an infinite time ($\Delta t = \infty$). Equation 9 is one form of the celebrated *uncertainty principle* or *principle* of *indeterminaucy*, which Warner Heisenberg first introduced in 1927. We shall explore its fuller meaning after we have given it another formulation.

For an object moving along the x-axis, the relation between the uncertainty Δp_x (magnitude) in momentum and the uncertainty $\Delta\lambda$ in wavelength can be obtained by taking the differential of Eq. (2).

$$\Delta \lambda = \frac{h}{p_x^2} \Delta p_x$$

Substituting this equation in Equation 7 and using Equation (2) give

$$\Delta\lambda\Delta x = \frac{h\Delta p_x}{p_x^2}\Delta x \geqslant \lambda^2$$

$$\Delta p_x \Delta x \geqslant \frac{(\lambda p_x)^2}{h}$$

$$\Delta p_x \Delta x \geqslant h \qquad \text{—(10)}$$

which is the other formulation of the Heisenberg uncertainty principle. It was assumed that the wavelength was measured over the finite distance Δx along the direction of wave propagation; with reference to the wave-particle duality, the quantity Δx can be interpreted as the uncertainty in the position of the particle. Thus, the formulation of the uncertainty relation given in Equation (10) says that the product of the uncertainties in position and momentum equals or exceeds Planck's constant. Note that the momentum and position referred to in the equation are both measured along the *same* direction; according to the uncertainty relation, it is impossible to specify simultaneously and with infinite precision the linear momentum and the *corresponding* position of a particle or photon. Although $\Delta p_x \Delta_x$ is equal to or greater than h, the product $\Delta p_y \Delta x$ can equal zero; there is no restriction on the simultaneous measurement of mutually perpendicular momentum and displacement.

The fundamental limitation on the certainty of measurements of energy and time or of position and momentum is compatible with the principle of complementarity. If the particle nature of, say, an electron is to be perfectly displayed, then both Δx and Δt must be zero. Therefore, when the particle aspect is chosen, the wave aspect is necessarily suppressed. All the quantities v, F, λ, and p are then completely uncertain, which follows either from the uncertainty principle or from complementarity. On the other hand, if the wave characteristics of a material particle or of electromagnetic radiation are to be defined perfectly, that is, if $\Delta v = 0$ and $\Delta\lambda = 0$ (also $\Delta E = 0$ and $\Delta p = 0$), then by the principle of complementarity or by the uncertainty principle we are prevented from giving simultaneously the distinctively particle characteristics of preeise location in space and in time, and x and t are completely uncertain.

Suppose that we want to represent an electron by its wave properties and yet localize it to some degree. We cannot use a single, sinusoidal wave; such a wave extends to infinity and is certainly not localized. We can, however, superpose several sinusoidal waves differing in frequency over a range of frequencies $\triangle v$ and have a *wave packet*. The component waves constructively interfere over a limited region $\triangle x$, identified as the somewhat uncertain localization of the "particle" and yield a resultant wave function ψ of the sort shown in Figure 3.2. Because there is necessarily a range in frequency $\triangle \nu$ and a range in wavelength $\triangle \lambda$, the associated momentum and energy are necessarily uncertain, and it is possible to predict precisely where or when the wave packet will reach another point and what the momentum and energy will then be.

Since the uncertainty relation implies an uncertainty in energy, of magnitude $h/\triangle t$ over time $\triangle t$, it also implies that the law of energy conservation may actually be violated—by that amount, $\triangle E = h/\triangle t$—but only for the time interval $\triangle t$. The greater the amount of energy borrowed or discarded, the shorter the time interval over which the nonconservation of energy exists.

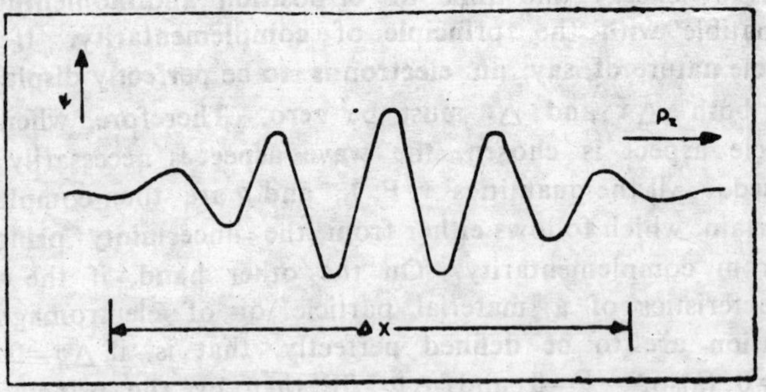

Fig. 3.2. The wave function of a wave pocket.

Similarly, the uncertainty in a particle's momentum $h/\triangle x$ implies that the law of momentum conservation may be violated, but only by the amount $\triangle p=h/\triangle x$ over a region $\triangle x$.

To derive the uncertainty principle in a different way, consider the diffracting of waves by a single, parallel-edged slit. A monochromatic plane wave is incident on a slit of width ω, and the diffraction pattern is formed on a distant screen, as shown in Figure 3.2. The location of the points of zero intensity is given by the equation.

$$\sin\theta=\frac{n\lambda}{\omega} \qquad \text{...(11)}$$

where λ is the wavelength and n is 1, 2, 3...The total intensity within the central maximum is much greater than the intensity within any of the secondary humps, since the area under itfar exceeds the area under any of the others. The area under the central hump is approximately three times the area under all the others; therefore, roughly three-fourths of the energy passing through the slit falls within this central region. The limits of the central region are given by Equation (11) with $n=1$:

$$\sin\theta=\pm\frac{\lambda}{\omega} \qquad \text{...(12)}$$

We have not yet specified what sort of wave is diffracted by the slit. If the wave consists of electromagnetic radiation, then the intensity of the diffraction pattern is proportional to E^2, the square of the electric field at the screen. If on the other hand, the wave consists of a beam of electrons, the intensity of the diffraction pattern is proportional to ψ^2, which is the square of the wave function at the screen and gives the probability of finding an electron at any point along the screen. Whether the waves are of electromagnetic radiation or of a material particle, the diffraction effects are pronounced only when their wavelength is comparable to the slit width for waves much shorter than the slit width, the intensity pattern on the screen reduces to a geometrical shadow cast by the edges of the slit.

Suppose now that we reduce drastically the amount of radiation or the number of electrons, as the case may be. Then when we observe the screen, we no longer see smooth variations along it but instead, photons or electrons arriving one by one. The intensity of the diffraction pattern is given by E for photons

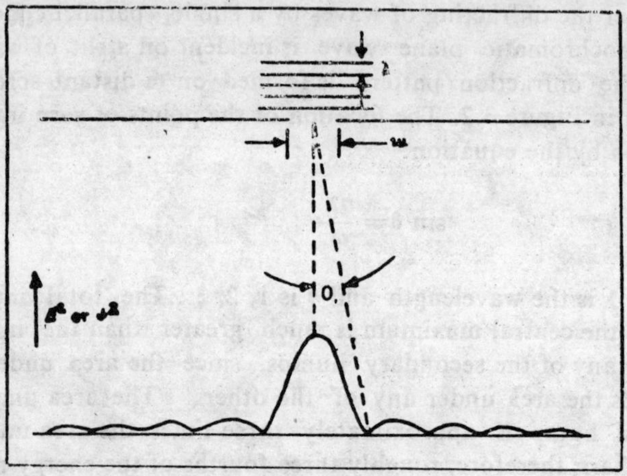

Fig. 3.3. Diffraction pattern of a monochromatic plane wave incident on a slit of width ω.

ψ^2 for electrons; thus, in Figure 3.3, the intensity represents the *probability* that a particle will strike a certain spot on the screen. There is a 75 per cent probability that it will fall within the central region in the diffraction pattern, a smaller probability that it will fall in any other, and *zero* probability that it will fall at the zeros in the pattern. At very low illumination, bright flashes appear over a large area of the screen. As time passes, more and more particles accumulate on the screen, and the distinct bright flashes merge and from the smoothly varying intensity pattern predict by wave theory.

There is no way of predicting in advance where any one electron or photon will fall on the screen. All that wave mechanics permits us to know is the probability that a particle will strike any one point. Before the particles pass through the slit, their momentum is known with complete precision both in magnitude (monochromatic waves) and in direction (vertically down in this case). When they pass through the slit, their position along a line in the x-direction, completely uncertain before they reached the slit, is now known, with an uncertainty $\Delta x = \omega$, the slit width. What is not known however, is precisely where any one particle will strike the screen. Any particle has approximately a three-to-one chance of falling anywhere within the central region, whose boundaries are given by Equation (12). There will be an uncertainty in the x-component of the momentum p that is *at least* as great as $p \sin \theta$; this can easily be seen in Figure 3.4. Therefore we write

$$\Delta p_x \geqslant p \sin \theta$$

Using Equation 12, we have

$$\Delta p_x \geqslant \frac{p\lambda}{\Delta x}$$

and since $p = h/\lambda$, we have

$$\Delta p_x \Delta x \geqslant h$$

the Heisenberg uncertainty relation.

Now suppose that Δx in our example is very large, that is, that the slit is very wide. Then the uncertainty in position is increased, and we cannot be certain where an electron is

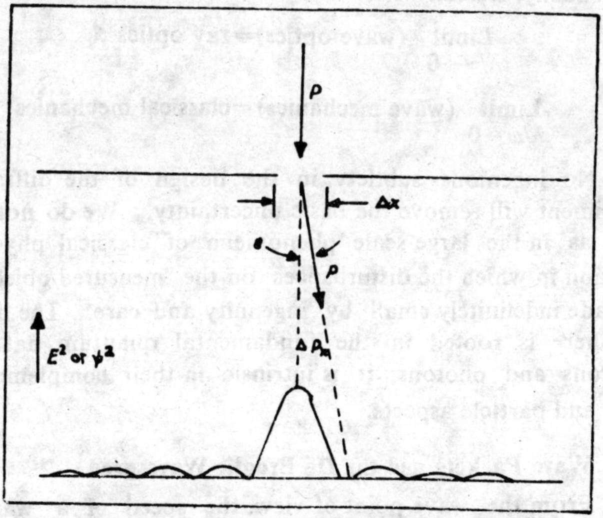

Fig. 3.4. An illustration of the uncertainty in the momentum of a particle that has passed through a single slit.

located along x. But the uncertainty in the momentum is reduced correspondingly; diffraction becomes less pronounced and essentially all electrons fall within the geometrical shadow. Centrarily, as the slit is reduced in width and Δx becomes very small, the diffraction pattern is expanded along the screen; and for the increase in our certainty of the electron's position, we must pay by a correspondingly greater uncertainty in the electron's momentum.

We see that when the slit width is much greater than the wavelength, particles pass undeviated through the slit to fall within the geometrical shadow. This agrees with classical mechanics, in which the wave aspect of material particles is ignored. Thus, there is a close parallel in the relation of wave optics to ray optics, and of wave mechanics to classical mechanics. Ray optics is a good approximation of wave optics whenever the wavelength is much less than the dimensions of obstacles or apertures that the light encounters; similarly, classical mechanics is a good approximation of wave mechanics whenever a particle's wavelength is much less than the dimensions of obstacles or apertures encountered by material particles. Symbolically, we can write

$$\lim_{\lambda/\omega \to 0} \text{(wave optics)} = \text{ray optics}$$

$$\lim_{\lambda/\omega \to 0} \text{(wave mechanics)} = \text{classical mechanics}$$

No ingenious subtlety in the design of the diffraction experiment will remove the basic uncertainty. We do not have here, as in the large-scale phenomena of classical physics, a situation in which the disturbances on the measured object can be made indefinitely small by ingenuity and care. The limitation here is rooted in the fundamental quantum nature of electrons and photons; it is intrinsic in their complementary wave and particle aspects.

3.4. Wave Packets and the De Broglie Wavespeed

From the wave point of view, the speed of a wave of frequency $\nu = E/h$ and wavelength $\lambda = h/d$ is given by

$$v_{r}h = v\lambda = \frac{E\,h}{h\,p} = \frac{E}{p} \qquad ...(13)$$

where the subscript denotes that the speed is the *phase* speed, the rate at which a point of constant phase travels through space, Using the angular frequency $\omega=2\pi\nu$ and the wave $k=2\pi/\lambda$, we can write

$$E = h\omega \qquad p = hk \qquad ...(14)$$

$$v_{ph} = \frac{\omega}{k}$$

where $h = h/2\pi$.

From the particle viewpoint, the relativistic energy E and momentum p of a particle of real mass Mo travelling at the speed v are given by

$$E = \gamma m_0 c^2 \qquad p = \gamma m_0\, v \qquad ...(15)$$

In assigning a speed v to a particle, we imply that its energy and momentum are to some degree localized and are transported at this speed. We can compare the particle speed ν with the phase speed v_{ph} of the associated wave by substituting Equation (15) in Equation (13)

$$v_{ph} = \frac{E}{p} = \frac{\gamma m_0 c^2}{\gamma m_0 v} = \frac{c^2}{v} \qquad ...(16)$$

First consider a photon, a particle with zero rest mass and speed $v=c$. From Equation (16), we have $v_{ph}=c$ for $m_0=0$. The associated electromagnetic wave travels at the same speed c as the photon does.

For a particle with a finite rest mass, the speed v must always be less than c, and Equation 16 becomes

$$v_{ph} > c \qquad \text{for } m_0 > 0$$

The phase speed of the associated wave of a nonzero rest mass particle must always *exceed* the speed of light. Consequently, a monochromatic wave associated with a material particle is unobservable -- a circumstance that is not disturbing for wave mechanics, since observations of ψ always involve the possibility

ψ^2 of finding a particle, not the speed with which the phase of ψ advances.

If we are to think of some object bearing energy and momentum as a particle, its energy and momentum confined to a small region of space, then the amplitude of the associated wave function must be concentrated in a relatively small region. The figure does not depict a wave of a single frequency, for such a wave would be strictly sinusoidal and have infinite extension ; it shows a wave packet, composed of waves of different frequencies. The component waves interfere constructively at the location of the particle, in the region in which the resultant wave function is large ; and they interfere destructively in all other regions. Since the particle's location is that for which the wave function and the probability are large, the speed v at which the particle moves is the speed at which the region of constructive interference advances through space.

When a group or packet, of individual sinusoidal waves differing in frequency *and in phase speed* combine to produce a region of storing constructive interference, the speed at which that region advances, the *group speed* v_{ge}, is related to the angular frequency ω and wave number k of the component waves by the relation.

$$v_{gr}=\frac{d\omega}{dk} \qquad \text{...(17)}$$

As the arguments have suggested, the speed v_{gr} of the *group* of the waves in the packet should equal the particle speed v. Let us prove it.

Since $\omega=E/h$ and $k=p/h$, we can write

$$v_{gr}=\frac{d\omega}{dk}=\frac{dE}{dp} \qquad \text{...(18)}$$

The particle's total energy E is related to its relativistic momentum p by

$$E=\sqrt{E_0^2+(pc)^2} \qquad \text{.. (19)}$$

Taking the derivative, we have

$$v_{gr} = \frac{dE}{dp} = \frac{pc^2}{\sqrt{E_0^2 + (pc)^2}} = \frac{pc^2}{E}$$

$$= \frac{\gamma m_0 v c^2}{\gamma m_0 c^2}$$

$$v_{gr} = v$$

The particle's speed is just the group speed of the particle's wave packet.

3.5. Quantum Description of a Confined Particle

For a particle that is completely free from any external infiuence, the classical description is that the particle moves in a straight line with a constant momentum (Newton's first law). In the language of wave mechanics, on the other hand a particle having a constant, well-defined momentum must be represented by a monochromatic sinusoidal wave with a well-defined wave-length. Therefore according to the uncertainly particle, the position of the particle is altogether uncertain and indeterminate. Most particles in nature, because of interactions with their surroundings, cannot be taken to be free, and a more general quantum law is needed. We shall now extend the quantum description to a confined particle.

Let us have the wave associated with a particle confined in the same way that a transverse wave on a string is confined. We assume that the particle moves freely back and forth along the x-axis but encounters an infinitely hard wall at $x=0$ and another at $x=L$, it is, then, confined between these boundaries The infinitely hard walls correspond to an infinite potential energy V for all values of x smaller than zero and larger than L. Because the particle is free between zero and L, its potential energy V in this region is constant. The situation we have described fits a particle in a one dimensional box, or a particle in an infinitely deep potential well. Because the walls are infinitely hard, the particle imparts non of its kinetic energy to them, its total energy remains constant, and it continues to bounce back and forth between the walls undeterred.

From the point of view of wave mechanics we can say that if the particle is confined within the limits states and its potential energy between zero and *L* taken as zero for convenience, then the probability of finding it outside these limits is zero. Therefore, the wave function ψ, whose square represents this probability, must be zero for $x \leqslant 0$ and $x \geqslant L$,

Mathemitically, the conditions of our problem are :

$$V=\infty \quad \text{for } x<0,\ x>L$$

$$V=0 \quad \text{for } 0<x<L$$

$$\psi=0 \quad \text{for } x\leqslant 0,\ x\geqslant L$$

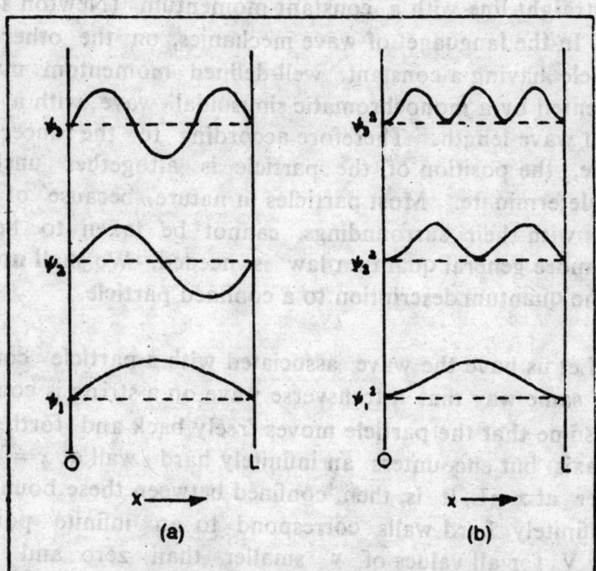

Fig. 3.5. The first three stationary states of particle in a one-dimensional box : (a) wave functions and (b) probability distributions.

Only the wave functions that satisfy these condition are allowed. Since the particle's momentum is constant in the entire region between the walls, it is represented, we know, by a sinusoidal wave. To satisfy the conditions at the boundarles, the only wavelengths that are allowed are those that will permit an integral number of half-wavelengths to be fitted between $x=0$ and $x=L$. The condition for the existence of *stationary*, or standing, waves is then

$$L=n\frac{\lambda}{2} \qquad ...(20)$$

where λ is the wavelength and n is the *quantum numbers* having the possible values 1, 2, 3,...

Fig. 3.5 shows the wave function ψ and the probability ψ^2 plotted against x, for the first three possible *stationary states* of the partcle in the box. Note that whareas ψ can be negative as well as positive, ψ^2 is always positive.

The probability distribution is such that ψ^2 is always zero at the boundaries. For the first state, $n=1$, the most probable location of the particle is the point midway between the two walls, at $x=L/2$; for the second state, $n=2$, however, the least probable location ($\psi^2=0$) is at $x=L/2$, which is to say that it is impossible for the particle to be located there.

Imposing the boundary conditions on ψ, that is, fitting the waves between the walls, has restricted the wavelength of the particle to the values given by Equation (20). If only certain wavelengths are permmitted, the magnitude of the momentum also is restricted to certain values, since $p=h/\lambda$. Therefore, the permitted momenta are given by

$$y=\frac{h}{\lambda}=\frac{hn}{2L} \qquad ...(21)$$

Finally, the kinetic energy E_k (and therefore the total energy E of the particle, since the potential energy is zero) is given by

$$E_k=E=\tfrac{1}{2}mv^2=\frac{p^2}{2m}=\frac{(hn/2L)^2}{2m}$$

$$E_n=n^2\frac{h^2}{8mL^2} \qquad ...(22)$$

where m is the particle's mass (this equation holds only for nonrelativistic speeds.) The subscript n signifies that the possible values of the energy depend only on the quantum number n (1, 2, 3,..) for fixed values of m and L.

It is convenient to write the possible energy values E_n in terms of the smallest value (for $n=1$). Since

$$E_l=\frac{h^2}{8mL^2} \quad \text{...(23)}$$

Equation (22) becomes

$$E_n=n^2E_1 \quad \text{...(24)}$$

The quantum state of lowest possible energy E_1 is called the *ground* state and corresponding energy E_1 the zero-point *energy.*

By Equations (21) and (22), we see that both the momentum (magnitude) and the energy of the particle in the one dimensional box are quantized. The first three allowed energies are shown in the energy-level diagram of Fig. 3.6. The particle cannot assume any momentum of energy except those particular momenta and energies that satisfy the boundary conditions placed on the wave function. The quantization of the energy is analogous to the classical quantization of the frequencies of waves on a string fixed at both ends.

Note that the particle's lowest possible energy (n=1) is not zero, but

$$E_1=\frac{h^2}{8mL^2} \quad \text{...(25)}$$

Although this result is surprising, it is an accord with the uncertainty principle, If the particle's energy were zero, with the particle then at rest somewhere within the box ($\Delta x=L$), both the momentum p and the uncertainty in the momentum Δp_x would be zero. But this would violate the uncertainty relation, since the product $\Delta p_x \, \Delta x$ would be (0) (L)=0, not h. Since the particle is known to be somewhere within the box, $\Delta x=L$, and by the uncertainty principle $\Delta p_x \geqslant h/\Delta x=h/L$.

For the lowest state. the particle's momentum p_x in one direction must be at least as great as the momentum uncer-

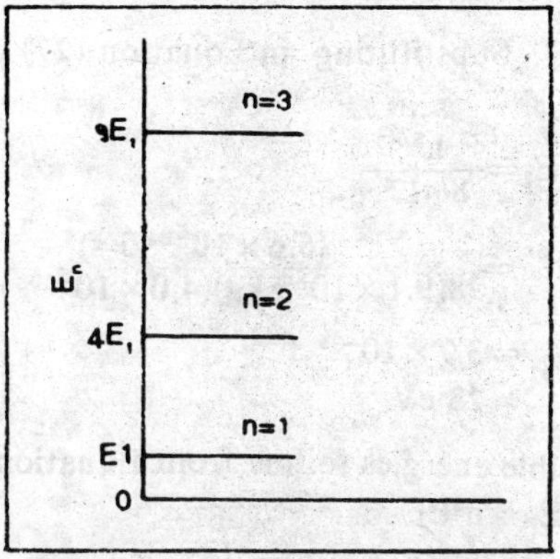

Fig. 3.6. First three allowed energies of a particle confined to a one dimensional box,

tainty Δp_x. Moreover, we cannot know whether the particle is travelling left or right, so that all told $\Delta p_x = 2p_x$. Under these circumstances, the particle's lowest energy is

$$E = \frac{p_x^2}{2m} = \frac{(\Delta p_x/2)^2}{2m} \quad \text{...(26)}$$

With $\Delta p_x = h/L$, we have $E = h^2/8mL$, exactly the energy of the first allowed state, given by Equation (23).

The problem of the particle in the box is somewhat artificial ; there is no such thing as an infinitely great potential energy, and a particle cannot be made completely free from all external influences while inside a box. Nevertheless, the problem is important because it reveals the quantization of the energy. Fundamentally, energy quantization occurs because only certain discrete of the wavelength can be fitted between the boundaries.

Example 3.1. Find the possible values of the energy and speed for the following particles confined to a one-dimensional box : (a) an electron of mass 9.1×10^{-31} kg constrained to move back and forth within an atomic distance $L = 4.0$ Å=

4.0×10^{-10} m ; (b) a relatively large object of mass 9.1 mg= 9.1×10^{-6} kg constrained to move along the x-axis with L= 4.0×10^{-2} m.

Sol. (a) Substituting in Equation (27) gives the ground state energy

$$E_1=\frac{h^2}{8mL^2}$$

$$=\frac{(6.6\times10^{-34}\ \text{J-s})^2}{8(9.1\times10^{-3}\ \text{kg})(4.0\times10^{-10}\ \text{m})^2}$$

$$=3.7\times10^{-19}\ \text{J}$$

$$=23\ \text{eV}$$

The next possible energies follow from Equation (21)

$$E_n=n^2E_1$$

or $$4E_1,\ 9E_1,\ 16E_1,\ldots\ldots$$

The permitted energies of the electron in a 4,0 Å box are shown on the energy level diagram of Fig 3.7. It is significant' is relation to atomic structure, that when an electron is confined to a distance approximately the diameter of an atom, its possible energies are in the range of a few electron volts, comparable to the binding energy of electrons in atom.

The electron's possible speeds can be found from Equation (21)

$$p_n=mv_n=\frac{nh}{9L}$$

or $$v_n=u\ \frac{h}{2mL}=nv_1$$

with the lowest speed v_1 given by

$$v_1=\frac{h}{2mL}$$

$$=\frac{6.6\times10^{-34}\ \text{J-s}}{2(9.1\times10^{-31}\ \text{kg})\ (4.0\times10^{-10}\text{m})}$$

$$=9{\cdot}1\times10^5\ \text{m/s}$$

Note that the ground-state speed is large, but still within the nonrelativistic region, since

$$\frac{v_1}{c}=\frac{9.1-10^5}{3\times10^8}=0.003$$

(b) Again, using Equation (23) for the relatively large object, we get

$$E_1=\frac{(6\,6\times10^{-3a}\ J,s)^3}{8(9.1\times10^{-6}\ kg)\ (4.0\times10\omega^2\ \ m)^2}$$
$$=2.3-10^{-41}\ eV$$

a fantastically small amount of energy.

The energy-level diagram for these circumstances, shown in Fig. 3.7(b), is plotted to the *same* scale as in Fig. 3.7(*a*). The spacing between adjacent energies for such macroscopic conditions is so very small that energy is effectively continuous. That is why we never see any obvious manifestation of the quantization of energy of macroscopic particle ; the quantization is there, but too fine to be discerned. This result agrees with the correspondence principle, which requires that the discrete energies of a bound system appear continuous in large scale phenomena.

The particle's speed follows from Equation (21)

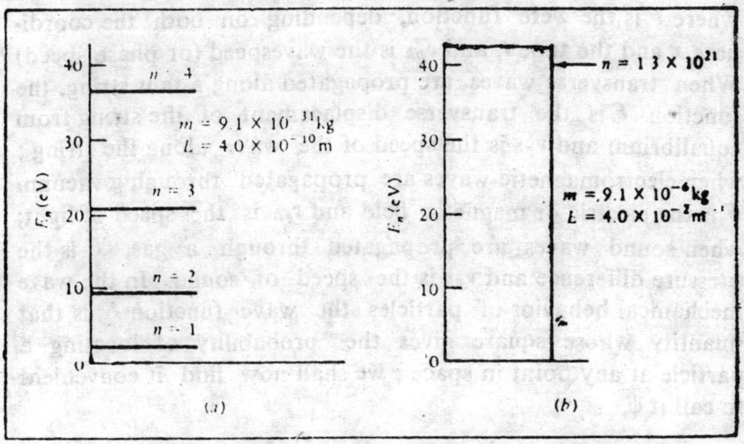

Fig. 3.7. (*a*) Allowed energies of an electron confined to a noe-dimensional box of atomic dimensions. (*b*) Allowed energies of a 9.1 mg particle confined to a 4-cm one- dimensional box.

$$v_n=\frac{p_n}{m}=n\ \frac{h}{2mL}$$

The lowest speed is

$$v_1=\frac{h}{2mL}$$

$$=\frac{(6.6\times 10^{-34}\text{ J.s})}{2(9.1\times 10^{-6}\text{ kg})\,(4.0\times 10^{-2}\text{ m})}$$

$$=9.1\times 10^{-28}\text{ m/s}$$

or a mere 10^{-7} Å per millennium. The particle is effectively at rest.

3.6. Schrodinger Equation

We now examine the quantum description of a bound particle in a potential-energy field V(x). From the classical one-dimensional wave equation, it is possible to arrive at the Schrodinger equation, whose solution yields the permitted were functions and quantized energies of bound systems.

Any wave propagated along the x-direction obeys the general wave equation,

$$\frac{\partial^2 F}{\partial x^2}=\frac{1}{v_{ph}^2}\frac{\partial^2 F}{\partial t^2} \qquad \text{...(27)}$$

where F is the were function, depending on both the coordinate x and the tune t, and v_{ph} is the wavespeed (or phase speed) When transverse waves are propagated along a taut string, the function F is the transverse displacement of ihe string from equalibrium and v_{ph} is the speed of the wave along the string ; when electromagnetic waves are propagated through a vacuum F is the electric or magnetic field and v_{ph} is the speed of light; when sound waves are propagated through a gas, F is the pressure difference and v_{ph} is the speed of sound. In the wave mechanical behavior of particles the wave function F is that quantity whose square gives the probability of locating a particle at any point in space ; we shall now find it convenient to call it ψ.

We consider only nonrelativistic systems whose total energy E is constant and whose particles move along the x-axis and are bound. Then, by the frequency $\nu=E/h$ associated with the bound particle is also constant. Assume that the wave function ($\psi(x, t)$ is separable into a spatial dependent term $\psi(x)$ and a time-dependent term f(t).

$$\psi(x, t)=\psi(x)\,f(t)$$

Since it is assumed that the frequency is precisely defined, the time dependent term $f(t)$ must very sinusoidally with time ; we can take it to be given by

$$f(t) = \cos 2\pi \nu t.$$

The second partial derivatives in Equation 27 then become

$$\frac{\partial^2 \psi}{\partial x^2} = f(i) \frac{d^2 \psi}{dx^2}$$

and $$\frac{\partial^2 \psi}{\partial t^2} = \psi(x) \frac{d^2 f}{dt^2} = -4\pi^2 \nu^2 f(t)\psi(x)$$

Setting these results in Equation 27 yields

$$f(t)\frac{d^2\psi}{dx^2} = -\frac{4\pi^2\nu^2}{v_{ph}^2} f(t)\psi(x)$$

$$\frac{d^2\psi}{dx^2} = -\left(\frac{2\pi}{\lambda}\right)^2 \psi = -\left(\frac{P}{h}\right)^2 \psi \quad \cdots (28)$$

where by Equation (13), the wavelength is $\lambda = v_{ph}/\nu$, and the momentum of the particle is $p = h/\lambda$.

We take the particle of mass m to be interacting with its surroundings (of infinite mass) through a potential-energy function $V = V(x)$. The total energy E of the system is then given by

$$E = E_k + V = \frac{p^2}{2m} + V$$

where E_k is the particle's kinetic energy. Then we have

$$p^2 = 2m(E - V)$$

and Equation (28) becomes

$$\frac{h^2}{5m}\frac{d^2\psi}{dx^2} + (E - V)\psi = 0 \quad \text{—(29)}$$

This equation is the one-dimensional, time-independent, non-relativistic Schrodinger equation. To get the Schrodinger equation that applies to particles moving in three dimensions, we merely replace $d^2\psi/dx^2$ in this equation with $\partial^2\psi/\partial x^2 + \partial\psi/\partial y^2 + \partial\psi/\partial z^2$.

In getting the Schrodinger equation, we assumed that the "particle" propagates as a wave; we used a wave equation. We also assumed, however, that it interacts with its surroundings as a particle; the potential energy V is given as a function of points in space. The complementarity principle is built into the Schrodinger equation.

If we know the force acting on a bound particle—that is, if we know the potential energy as a function $V(x)$ of the particle's position—we can find the allowed wave functions and allowed energies of the system. Because the square of the wave function is proportional to the probability of finding the particle at position x, an acceptable solution $\psi(x)$ must be finite, continuous, and single-valued, and most especially, it must be consistent with the boundary conditions imposed by the character of the potential energy V. Indeed, it is imposing boundary conditions on the wave function that leads to the quantization of the energy of a bound system. In nonanalytical terms, we must regard the particle as a wave reflected back and forth within the confines of a bound system, forming standing waves; it is fitting stationary waves to boundary conditions that leads to the quantized values of a system's permitted energies. If the potential energy depends on position, but the system's total energy E is constant, then the particle's kinetic energy, momentum and wavelength all must depend on position. More specifically, the wavelength is given by

$$\lambda = \frac{h}{p} = \frac{h}{\sqrt{2m(E-V)}} \qquad \text{—(30)}$$

For any given energy state, the energy E (and frequency ν) is constant, but the wavelength λ (and momentum p) is not; λ and p vary with position x according to Equation (30). Some examples of the application of Equation (29) follow.

3·7. Particle in a One-Dimensional Potential Well of Infinite Height

The simplest problem to solve with the Schrodinger equation

concerns a single particle confined to a one-dimensional box of width L; that is, the particle is imagined to be in an infinitely deep potential well, as shown in Figure 3.8. The potential energy $V(x)$ is constant zero within the box and rises to infinity at the boundaries, $x=0$ and $x=L$. For the interval $0<x<L$, the Schrodinger equation, Equation (29), then becomes, with $V=0$.

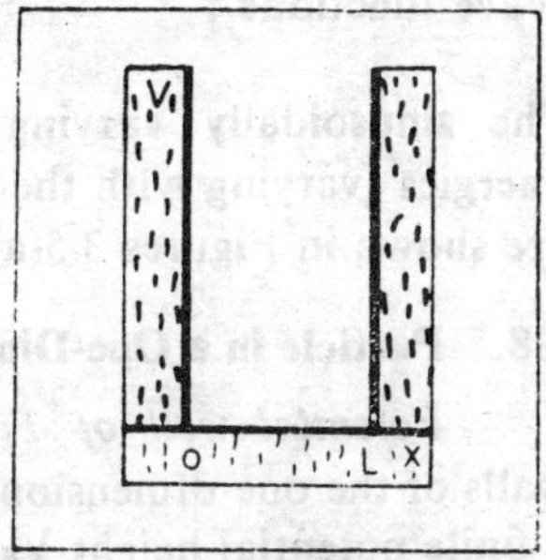

Fig. 3.8. Potential energy of an infinitely deep well of width L.

$$\frac{h^2}{2m}\frac{d^2\psi}{dx^2}+E\psi=0$$

or
$$\frac{d^2\psi}{dx^2}=-B^2\psi \qquad -(31)$$

where $B^2=2mE/h^2$. Since the walls are infinitely high, the particle cannot be found outside the box. This implies that $\psi(x)$ must be zero for all points at the boundaries of the box and beyond. Thus, the allowed solutions must be consistent with the boundary conditions, which are $\psi(0)=0$ and $\psi(L)=0$. A suitable solution is

$$\psi(x)=A\sin Bx$$

which can be verified by substitution in Equation (31). The first boundary condition $\psi(0)=0$ is satisfied for the sine function; it would *not* be satisfied for a cosine function, so the cosine solution is discarded. The second boundary condition $\psi(L)=0$ is satisfied only if $BL=n\pi$, where n is an integer (since $\sin n\pi=0$)' Substituting for B, we then have

$$n\pi=BL=\sqrt{2mE}\frac{L}{h} \qquad -(32)$$

The energies and wave functions of a free particle confined to a box of infinitely high walls are therefore

Energies :
$$E_n=\frac{n^2\pi^2h^3}{2mL^2}$$

Wave functions : $$\psi_n(x) = A_n \sin \frac{n\pi x}{L} \qquad \text{—(35)}$$

The sinusoidally varying wave functions and the quantized energies (varying with the square of the quantum number n) are shown in Figures 3.5(a) and 3.6.

3.8. Particle in a One-Dimensional

Potential well of Finite Height. Suppose now that the walls of the one-dimensional box are not infinitely high but have a finite potential height V_d, as shown in Figure 3.9. Again V is taken to be zero at the bottom of the well, but the zero of x is now taken to be at the center of the well. We know that in classical physics the particle must remain inside the square well if its kinetic energy is less than the magnitude of V_d; that is, if the particle is bound, with a total energy E less than V_d, it

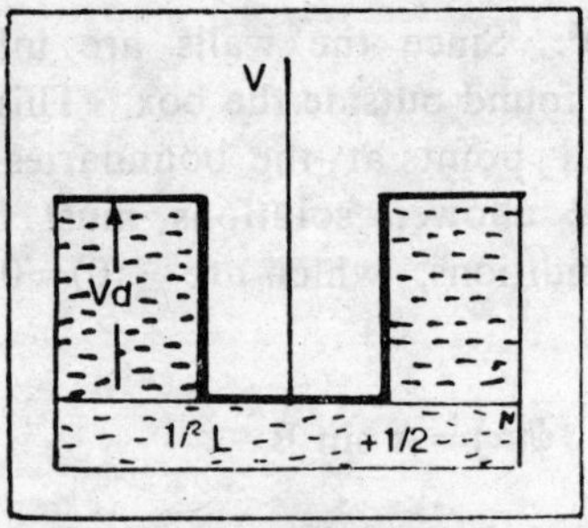

Fig. 3.9. Potential energy of a well of finite depth V_d and width L.

cannot possibly be found anywhere but in the region $-L/2 < x < L/2$. In wave mechanics the condition is less stringent; the particle *may* be found outside the limits defined by classical physics. What appears superficially to be a violation of energy conservation is permitted through the uncertainty relation. Thus, the wave function may actually be nonzero outside the well, that is, at $x < -L/2$ and $x > L/2$. In the interior of the well, where the potential energy is constant, the wave functions are sinusoidal, as before, but they now do not necessarily drop to zero at the boundaries, $x = -L/2$ and

$x=+L/2$. Outside the well, where $x<-L/2$ or $x>L/2$, the Schrodinger equation (Equation 29) becomes :

$$\frac{d^2\psi}{dx^2}=\frac{2m(V_d-E)}{h^2}\psi \qquad -(36)$$

The potential energy of a bound particle exceeds the total E, so that V_d-E, appearing on the right-hand side of the equation, is *positive*. If we use $C^2=2m\,(V_d-E)/h^2$, this equation becomes

$$\frac{d^2\psi}{xd^2}=C^2\psi \qquad -(37)$$

where C^2 is a positive quantity for the regions outside the potential well.

The possible solutions of this equation are $\psi_+=A_+e^{+cx}$ and $\psi_-=A_-e^{-cx}$, which can be verified by substitution. First we concentrate on the exterior region on the right. where $x>L/2$. Here x is always positive and we can rule out $\psi_+=A_+e^{+cx}$ on the following grounds. Although an acceptable wave function may be nonzero beyond the walls of the one-dimensional box, it cannot be infinite; and if ψ_+ where a solution, we should find it to be infinite at infinite distances from the box along the positive of the x-axis, implying an infinite probability of finding the particle at the greatest distances from the box. Clearly, this is physically untenable and we are then left with the solution $\psi_-=A_-e^{-cx}$, which means that the wave function decays exponentially in a direction away from the box on the positive of the x-axis and becomes zero at infinitely great distances. Similar arguments show that the wave function also decays exponentially along the negative of the x-axis, since x is always negative to the left of the box, the appropriate form there is $\psi^+=A_+e^{+cx}$.

To sum up. the wave function is a sinusoid inside the box and a decaying exponential outside it. The inside and outside wave functions must, however, join smoothly at the points $x=+L/2$ and $x=-L/2$. At both these points

$$\psi_{inside}=\psi_{outside} \qquad -(38)$$

$$\frac{d\psi_{inside}}{dx} = \frac{d\psi_{outside}}{dx} \quad (39)$$

The complete wave functions of the first two allowed states for particular values of V_d and L are shown in Fig. 3.10

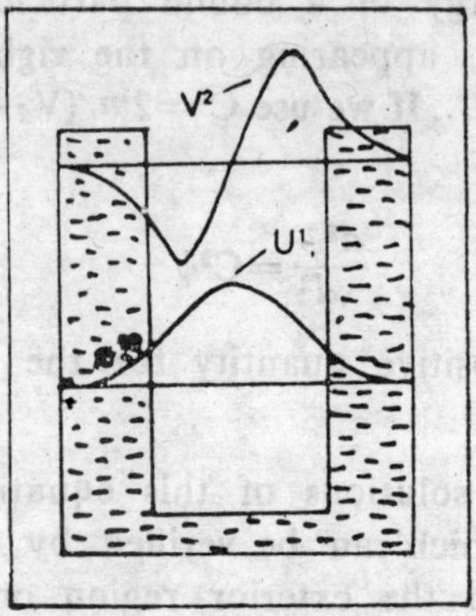

Fig. 3.10, Wave functions for the first two states of a partiele in a potential energy well of finite depth.

The corresponding allowed energy states are somewhat lower than the allowed energy states for a box with infinitely high walls. This is a consequence of a greater wavelength inside the box—greater since the wave functions do not drop to zero at the walls; hence there is a lower momentum, a lower kinetic energy, and a lower total energy. Furthermore, for a finite potential well V_d, the number of bound energy states is **finite.** Bound states can exist only for $E < V_d$.

3.9. Tunnel Effect

Since according to wave mechanics, a particle can spill over its classical confines, we have a curious possibility when a particle encounters a potential wall of finite width as well a height (see Fig. 3.11). In classical physics, a particle with too little kinetic energy to get over the wall would never get through to the other side by either climbing or passing through; it simply could not, because of the conservation of energy. In wave mechanics getting to the other side is a possibility. As we have seen, the wave function is nonzero and decays exponentially when the potential energy exceeds the total energy. Thus, the wave function of a particle approaching the wall from the left is sinusoidal to the left of the wall, exponential through the wall, and sinusoidal, but of much smaller amplitude, to the right of the wall. Since the amplitude of the wave funetion gives a measure of the probability of locating a particle, there is a small but finite probability that the particle that has approached from the left will be on the *right*. In other words, there is a high probability that the particle will be found on the left, a smaller probability that it will be found within the wall, and a still probability that it will be found on the right. The particle, or more properly, the wave, can penetrate, or tunnel through, the classically insurmountable barrier. The probability

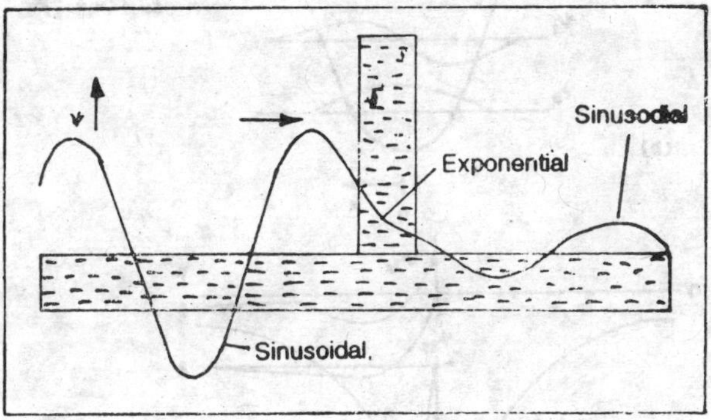

Fig. 3.11. **Wave function of a particle incident from the left upon a barrier of finite height and width.**

of this *tunnel* effect is vanishingly small except at the atomic and nuclear level. The effect is observed, however, in the behavior of certain semiconducting devices (the so-called tunnel diodes) and in the emission of a particles from unstable. heavy nuclei.

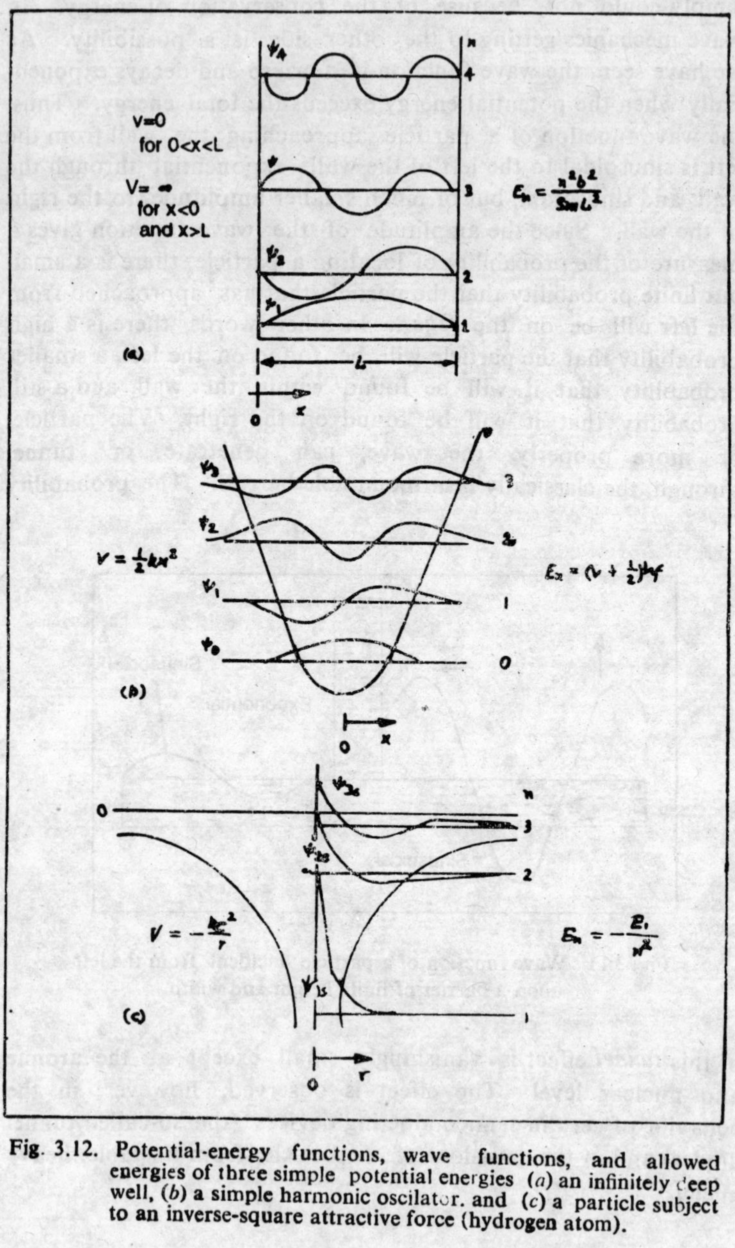

Fig. 3.12. Potential-energy functions, wave functions, and allowed energies of three simple potential energies (*a*) an infinitely deep well, (*b*) a simple harmonic oscilator. and (*c*) a particle subject to an inverse-square attractive force (hydrogen atom).

3.10. Hydrogen Atom and Its Wave Functions from Schrodinger Equation

The problem is finding allowed wave functions of the time-independent (nonrelativistic) Schrodinger equation,

$$\frac{\partial^2\psi}{\partial x^2}+\frac{\partial^2\psi}{\partial x^2}+\frac{\partial^2\psi}{\partial z^2}+\frac{2m}{h^2}(E-V)\psi=0) \quad ...(40)$$

for a particle of mass m (the electron) under the influence of an inverse-square attractive force (the Coulomb force of the nucleus).

We wish to find solutions to the Schrodinger equation for the Coulomb electric potential energy.

$$V=-\frac{ke^2}{r} \quad ...(41)$$

between two point charges, each of magnitude e, separated by r. Again, the positive point charge is assumed to remain fixed at the origin. To arrive at the most general solutions of the Schrodinger equation, we might transform the rectangular coordinates of Equation (40) into spherical co-ordinates (r, θ, ϕ). We should thereby be expecting the solutions to take on their simplest mathematical form when the co-ordinate system matches the potential energy (involving the co-ordinate r only). We shall, however, follow a still simpler course.

Since the potential energy depends only on the radial distance r, there must be a class of wave functions satisfying Equation (40) that are spherically symmetrical, that is, that depend only on the co-ordinate r.

$$\psi=F(r)$$

As written in Equation (40), the Schrodinger equation contains partial derivatives involving the rectangular co-ordinates x, y, and z. Since we seek solutions depending only on r, we must first transform this equation into one involving derivatives of r only. Clearly,

$$r=\sqrt{x^2+y^2+z^2}$$

Thus,

$$\frac{\partial r}{\partial x}=\frac{1}{2}\left(\frac{2x}{(x^2+y^2+z^2)^{1/2}}\right)=\frac{x}{r}$$

Now consider $\partial^2\psi/\partial x^2$. Remembering that ψ is to depend only on on r, we have

$$\frac{\partial\psi}{\partial x}=\frac{d\psi}{dr}\frac{\partial r}{\partial x}=\frac{x}{r}\frac{d\psi}{dr}$$

The second derivative is then

$$\frac{\partial^2\psi}{\partial x^2}=\frac{\partial}{\partial x}\left(\frac{\partial\psi}{\partial x}\right)=\frac{\partial}{\partial x}\left(\frac{x}{r}\frac{d\psi}{dr}\right)$$

$$=\frac{1}{r}\frac{d\psi}{dr}\left(\frac{\partial x}{\partial x}\right)+x\frac{d\psi}{dr}\frac{\partial}{\partial x}\left(\frac{1}{r}\right)+\frac{x}{r}\frac{\partial}{\partial_x}\left(\frac{d\psi}{dr}\right) \quad \text{--(42)}$$

Therefore, $$\frac{\partial^2\psi}{\partial x^2}=\frac{1}{r}\frac{d\psi}{dr}-\frac{x^2}{r}\frac{T\psi}{dr}+\frac{x^2}{r}\frac{d^2\psi}{dr^2} \quad \text{...(43)}$$

The derivatives for $\partial^2\psi/\partial y^2$ and $\partial^2\psi/\partial z^2$ are just as given here except that y and z replace x. Adding the three equations, we have

$$\frac{\partial^2\psi}{\delta x^2}+\frac{\partial^2\psi}{\partial y^2}+\frac{\partial^2\psi}{\partial z^2}=\frac{3}{v}\frac{d\psi}{dr}-\left(\frac{x^2+y^2+z^2}{r^3}\right)\frac{d\psi}{dr}+\left(\frac{x^2+y^2+z^2}{r^2}\right)\frac{d^2\psi}{dr^2}$$

$$=\frac{2}{r}\frac{d\psi}{dr}+\frac{d^2\psi}{dr^2}$$

Equation (40) then becomes

$$\frac{d^2\psi}{dr^2}+\frac{2}{r}\frac{d\psi}{dr}+\frac{2m}{h^2}\left(E+\frac{ke^2}{r}\right)\psi$$

$$=0 \quad \text{.. (44)}$$

Rather than try to solve this analytically in detail, we shall choose what seems to be a reasonable form of the wave function for the ground state and then test whether it is a solution. We know that the electron is more likely to be found near the nucleus rather than very far from it and the probability of finding the electron at infinity is zero. This implies that the wave function ψ is relatively large for small r and is zero for $r \to \infty$. A wave function satisfying these requirements is

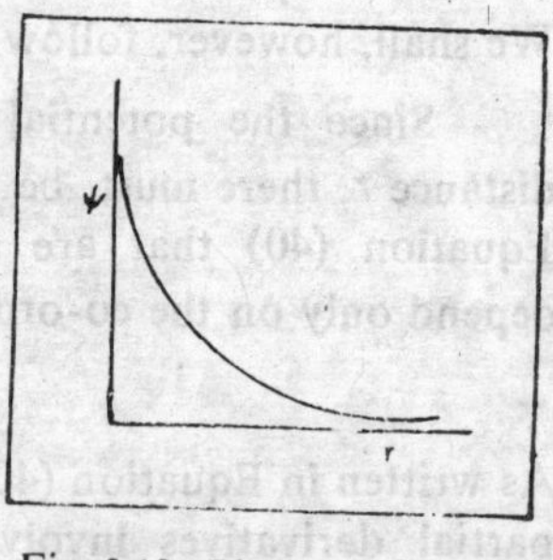

Fig. 3.13. The wave function $\psi=e^{-r/a}$ for the ground state of hydrogen.

$$\psi = e^{-m} \qquad ...(45)$$

where the size of a determines the exponential decay outward along r, see Fig. (3.13). Now we test whether this trial wave function satisfies the Schrodlnger equation (43).

From Equation (45) we have

$$\frac{d\psi}{dr} = -ae^{-ra} = -a\psi$$

$$\frac{d^2\psi}{dr^2} = a^2e^{-ra} = a^2\psi$$

Setting ψ, $d\psi/dr$, and $d^2\psi/dr^2$ in Equation 46, we have

$$\left[a^2 - \frac{2}{r}a + \frac{2m}{h^2}\left(E + \frac{ke^2}{r}\right)\right]\psi = 0 \qquad ...(46)$$

One possible solution is $\psi = 0$ everywhere, but this is unacceptable. A nonzero solution for ψ occurs when the quantity within the brackets equals zero. Regrouping the terms in Equation 47 gives

$$\left(a^2 + \frac{2m}{h^2}E\right) + \frac{1}{r}\left(\frac{2mke^2}{h^2} - 2a\right) = 0 \qquad ...(47)$$

This equation must hold for all values of r from zero to infinity. For r small, $1/r$ is large ; therefore, the second parenthetical term in Equation 47 must equal zero. For r large, $1/r$ is small, and the first term must equal zero. With the second term equal to zero, we have

$$a = \frac{mke^2}{h^2} \qquad ...(48)$$

With the first term in Equation 47 zero,

$$E = -\frac{a^2h^2}{h^2} = -\left(\frac{mke^2}{h^2}\right)^2\frac{h^2}{2m}$$

$$E = -\frac{mk^2e^4}{2h^2} = -13.6 \text{ eV} \qquad ...(49)$$

The energy of the atom is $E = -13.6$ eV, just the energy of the hydrogen ground state, according to the Bohr theory.

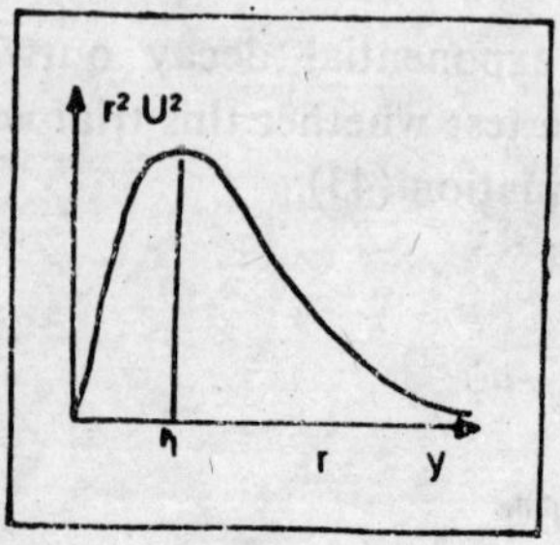

Fig. 3.14. The probability of finding an electron in the hydrogen ground state between r and $r+dr$ (proportional to $r^2\psi r^2$) as a function of distance r from the nucleus

From the wave description, Equation (45) of the hydrogen atom in its ground-state energy, what can we infer about the size of the atom ? Recall that $\psi^2\ dv$ is the probability of finding a particle in the volume element dv. For a hydrogen atom in the ground state, the probability of the electron's being within a small volume element dv is proportional to

$$\psi^2\ dv = e^{-2ra}\ dv$$

For this state, we are more likely to find the electron in a volume element dv at the nucleus ($r=0$) than in the *same* volume element anywhere else. We now ask what the probability is of the electron's being *between* r and $r+dr$, that is, within a spherical shell of radius r, thickness dr, thickness dr, and volume $4\pi r^2\ dr$.

For *equal volumes dv*, the probability $\psi^2\ dv$ is the maximum at the origin. For spherical shells of *equal thickness dr*, however, the volume $dv = 4\pi r^2\ dr$ of the shell is very small at the origin and very large at large distances from the origin ; therefore, there must be a maximum probability of finding the lectron in the spherical shell between $r=0$ and $r=\infty$.

For a shell of radius r and thickness dr, the volume element is $dv = 4\pi r^2\ dr$, the corresponding probability density is proportional to $r^2 e^{-2ra}$. This is shown plotted in Fig. 8.14. The curve rises initially (because of r^2), reaches the maximum, and then goes to zero for large r (because of ϵ^{-2ra}). At what distance r from the nucleus, within the range dr, is the electron most likely to be located ? This value of r is, of course, the value corresponding to the maximum in Fig. 6.16. The peak of $r^2 e^{-2ra}$ occurs at the following location :

$$\frac{d}{dr}(r^2 e^{-2ra}) = 0$$

$$2re^{-2ra}-2ar^2e^{-2ra}=0$$

$$r_{max}=\frac{1}{a}=\frac{h^2}{mke^2}$$

From this equation it follows that the maximum is exactly at the radius of the first Bohr orbit. Although the electron may be found anywhere, when in the ground state it is more likely to be a distance r_1 from the nucleus than any other distance from it.

Still other spherically symmetrical wave functions exist. It is easy to show by substitution in Equation (44) that the wave function $\psi_2=e^{-ra/2}(2-ra)$ also is a solution. The corresponding energy is found to be $E_2=-13.6\text{ eV}/4=-3.4$ eV, the energy of the first excited state of hydrogen. Indeed, the energies E_n of the hydrogen atom for all spherically symmetrical wave functions are given by

$$E_n=-\left(\frac{mk\,e^4}{2h^2}\right)\frac{1}{n^2}=-\frac{E_1}{n^2} \quad \text{...(50)}$$

with $n=1, 2, 3, \ldots$, again agreeing with the Bohr theory.

Since the potential energy $V=-ke^2/r$ is spherically symmetrical, there are spherically symmetrical wave functions, as we have seen. But there are also nonspherically symmetrical wave functions. The simplest of these is

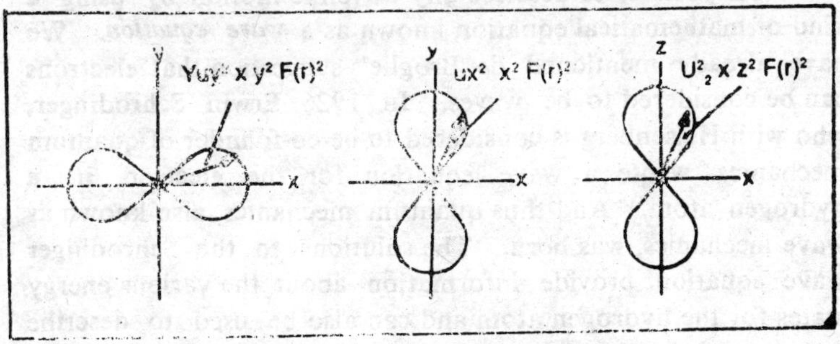

Fig. 3.15. Polar plot of the variation of ψ^2 with angle for p states.

$$\psi=xF(r) \quad \text{.. (51)}$$

where F(r) is again a function of r only. That this is a solution is tested by computing the derivatives $d\psi/dr$ and $d^2\psi/dr^2$ and setting them in Equation 44.3. The corresponding energy with F(r)= e^{-0/r^2} is found to be $E_2 = -13.6$ eV/4 = -3.4 eV, that of the first excited state of hydrogen.

If $\psi = x$F(r) is a solution, then so are $\psi = y$F(r) and $\psi = z$F(r), where F(r) is the same for all three states. The energy is the *same* for all three states. The energy is the *same* for all three states. Nonspherically symmetrical wave functions of the type

$$\psi = x\text{F}(r)$$

$$\psi = y\text{F}(r)$$

and

$$\psi = z\text{F}(r)$$

are called p states (the spherically symmetrical solutions are called s states). Fig. (3.15) shows probability distributions for a p state (or a p orbital) in a polar diagram.

All spherically symmetrical states have in common that the atom's orbital angular momentum is zero. For nonspherically symmetrical states, such as the p states of Fig. (3.15) it is *not* zero ; indeed, the orbital angular momentum (along any one direction in space) can be shown always to be an integral multiple of h.

3.11. Shapes of Atomic Orbitals

It is possible to describe any wavelike motion by using a kind of mathematical equation known as a *wave equation.* We have already mentioned de Broglie's suggestion that electrons can be considered to be waves. In 1926 Erwin Schrodinger, who with Heisenberg is considered to be co-founder of quantum mechanics, wrote a wave equation for the electron in a hydrogen atom. And thus quantum mechanics, also known as wave mechanics, was born. The solutions to the Schrodinger wave equation provide information about the various energy states for the hydrogen atom and can also be used to describe other atoms. The approach can even be used to descripe bonding and the properties of molecules. Because quantum-mechanical calculations are difficult and tedious to perform with accuracy, many could not be done until the advent of the high-speed electronic computer, but today the practical power of the quantum-mechanical approach has finally been realized.

A wave equation is not much like an algebraic equation. It is of a type known as a *differential* equation. It is important to know that a differential equation normally has not just one, or two, but rather a whole series of solutions. When, for example, a wave equation is written and then solved for the case of a vibrating string, each solution corresponds to one mode of vibration of the string and can be used to find wavelength, frequency, energy, etc., for that mode. This is also true for wave equations written for electrons in atoms. From such a wave equation a series of separate solutions is obtained, each of which corresponds to a different energy state. Thus in quantum mechanics it is not necessary to *assume* quantization of energy, as it was in the case of the Bohr atom. Instead, the condition of energy quantization falls quite naturally out of the theory. It happens as a direct consequence of the electron's wavelike character.

Each solution to a wave equation is called a *wave function* and is represented by the Greek letter ψ. (This should be pronounced "psee," but is very commonly mispronounced "psi" or "si," to rhyme with "eye.") The first solution to the wave equation for the hydrogen atom can be written as

$$\psi_{1s} A e^{-r} \qquad \text{—(52)}$$

Here A and B are constants {which can be evaluated from the theory), *e* is also a constant, being the base of the natural logarithms (2.71828 . . .), and *r* represents the distance from the nucleus. ψ corresponds to the *amplitude* of the electron wave. Most important is the value of ψ^2. This represents the probability of finding the electron within some tiny region which is at a distance *r* from the nucleus.

We now see how quantum mechanics meets the challenge posed by the uncertainty principle. Instead of providing a concrete picture of an electron's position and motion, it speaks in terms of probabilities of finding the electron at various places. Since ψ^2 is a probability of finding an electron within some tiny region or volume, it is a probability per unit volume and is called, therefore, a ***probability density*** (recall that ordinary density is mass per unity volume).

Is Orbital

Look at Eq. (52) again. It is the solution to the wave equation for an electron in a 1*s* orbital. The exponent is $-Br$, and because of the negative sign, ψ will decrease (quite rapidly) as *r* increases. This means that the farther we go from the nucleus, the smaller will be the probability of finding the electron. This relation between probability density, ψ^2, and distance from the nucleus, *r*, can be shown on a graph, Fig. 3.16, in which values of ψ^2, are plotted against *r*. Looking at the graph [or at Eq. 52] we see that the probability is highest at (immediately adjacent to) the nucleus. This does *not* mean that the electron *is* next to the

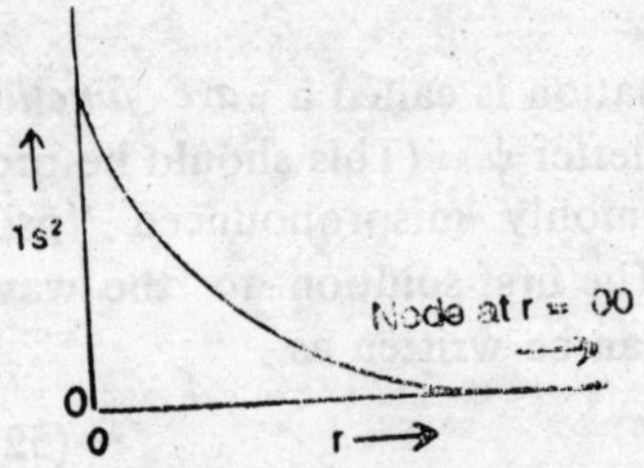

Fig. 3.16. Probability-density plot for a 1*s* electron. This shows the probability of finding a 1*s* electrnn in a tiny volume at a distance *r* from the nucleus.

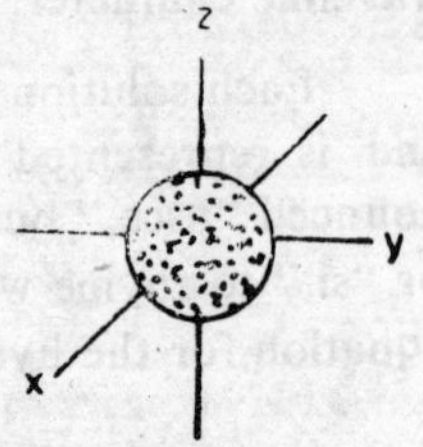

Fig. 3.17. The electronic charge cloud of a 1*s* electron.

nucleus, merely that the chances of finding it are highest there. The chance of finding the electron gets smaller the farther we go from the nucleus, but that chance never really gets to zero

(except at infinity). It soon gets so low, however, that for practical purposes we need not worry about it.

Thinking in terms of probabilities like this may seem difficult at first. Perhaps the following discussion will make it easier. Imagine that the electron is rapidly moving in some way which is pretty much undescribable, but in which it spends more time near the nucleus and less time at distances farther and farther away. In this way the negative charge carried by the electron becomes, in effect, smeared out into a blob or cloud which is most concentrated or dense at the center and which becomes progressively thinner going away from the nucleus. This electron *charge cloud* takes the shape of a sphere (for a 1*s* electron) and would look, if we could see it, something like Fig. 3.17. Now suppose that we cut the sphere through the middle and look at a cross seetion. We would see something like Fig. 3.18. The intensity of shading is shown greatest at the nucieus ; this indicates that the density or concentration of electronic charge is greatest there. In other words the probability of finding the electron is greatest at the nucleus, which is exactly what Eq. (52) and Fig. 3.16 say.

Another way of depicting the probability-density distribution is obtained by constructing a surface (in three dimensions), every point on which has the same value of ψ^2. For the electron in a 1*s* orbital such a surface is a sphere, as shown in Fig. 3.19. Since the actual shape of the electron charge cloud is spherically symmetrical, and number of spheres could be so constructed, the larger ones containing a larger percentage of

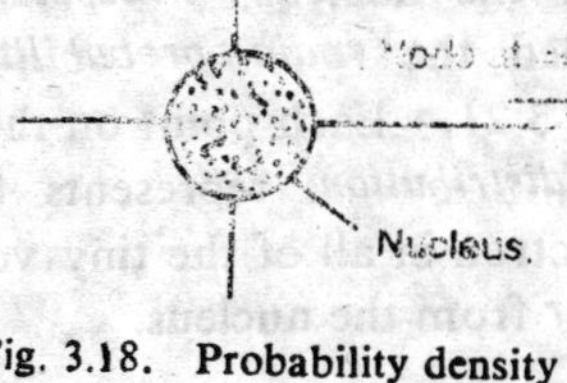

Fig. 3.18. Probability density for a 1*s* electron. This is a cross sectional view of the charge cloud shown in Fig. 6.14.

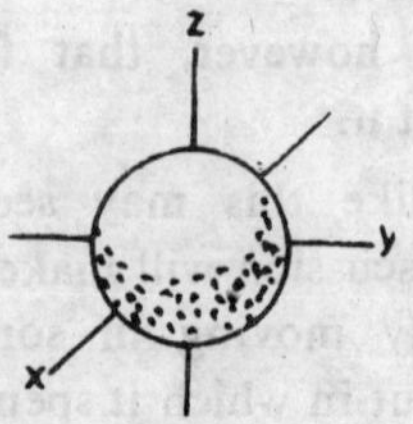

Fig. 3.19. Boundary surface for a 1*s* orbital.

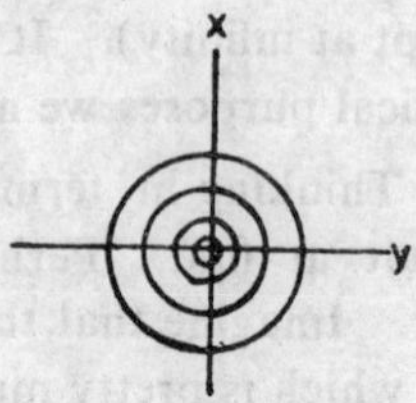

Fig. 3.2. Contour diagram for a 1*s* electron.

the total electronic charge. These surfaces of constant probability density, ψ^2, are called *boundary surfaces* and are usually constructed large enough so that they enclose a large percentage, say 90 per cent, of the electronic charge. Stated differently, the probability that the electron is *somewhere inside* the sphere is 90 per cent.

A last way of showing the shape of a probability-density distribution is shown in Fig. 3.20. Each circle in the diagram is the cross-section of a boundary surface, the larger ones obviously enclosing more of the electronic charge than the smaller. Such a diagram is often called a *contour diagram* because each line represents a contour of constant probability density in much the same way as lines on a topographic map represent contours of constant elevation.

ψ^2 tells us the probability of finding an electron within a tiny element of volume *in a specified location*. It is also common to show the probability of finding an electron *at a specified distance* from the nucleus, *irrespective of angular direction*. This is called the *radial probability* and is plotted for a 1*s* electron in Fig. 3 21. Each point on the curve, called the *radial probability distribution*, represents the total probability of finding an electron in all of the tiny volume elements which are at a distance *r* from the nucleus.

Notice that the radial probability distribution curve starts at 0 (at $r=0$), increases, goes through a maximum, and then decreases. Why is this curve different from the probability-density curve shown in Fig. 3.16. ? The answer is found in the fact that the number of tiny volume elements at a distance r from the nucleus increases as r increases. This is true because all these volume elements are part of a thin spherical shell of radius r, and as r increases, the number of volume elements out of which the shell is composed effectively increases. At small distances from the nucleus the radial probability is low even though ψ^2 is high, because the number of volume elements in the shell is low. As r increases, the number of volume elements

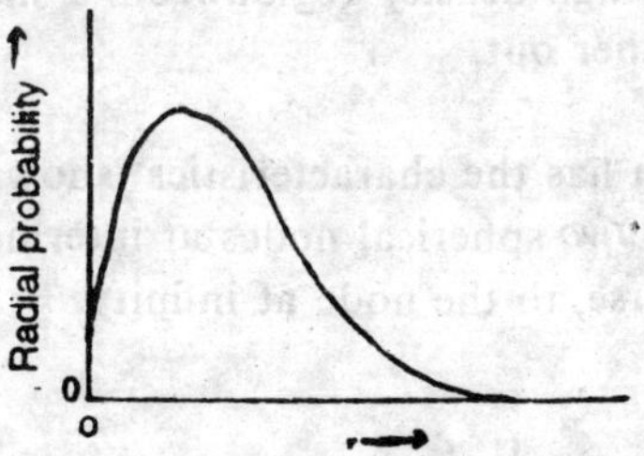

Fig. 3.21. Radial probability distribution for a 1*s* electron.

increases while ψ^2 decreases. This makes the radial probability, that is, the probability of finding the electron within the thin spherical shell, first increase, as the size of the shell increases while ψ^2 is still fairly high, and then decrease, as ψ^2 drops to very low values. (*Note : within the shell* means "in the shell," *not* enclosed by it. Think of the tiny volume elements as being like microscopic shreds of tomato skin, not like pieces of the interior of the tomato.)

The maximum in the 1*s* radial probability curve shown in Fig. 6.18 shows the distance from the nucleus at which the electron is most probably found. This means that it is most probably found somewhere in a thin, spherical shell of this radius. It is interesting to note that this distance is exactly that predicted by the Bohr theory for the radius of the orbit of a 1*s* electron.

The 2s and 3s Orbitals

Each *s* orbital is spherically symmetrical. In addition each has a spherical *node* at infinity. (This statement merely means that the probability of finding an electron approaches zero at a distance approaching infinity in any direction from the nucleus.) A 2*s* orbital has, in addition, another spherical node. This second node shows up as a place where the value of ψ briefly reaches zero in a probability-density plot (Fig. 3.22). Also shown is a cross section of the electron charge-cloud representation of a 2*s* electron. The node is a spherical surface separating the central, high-density region from a shell-shaped high-density region farther out.

A 3*s* electron has the characteristics shown in Fig. 3.23. Notice that it has *two* spherical nodes at intermediate distances in addition, of course, to the node at infinity.

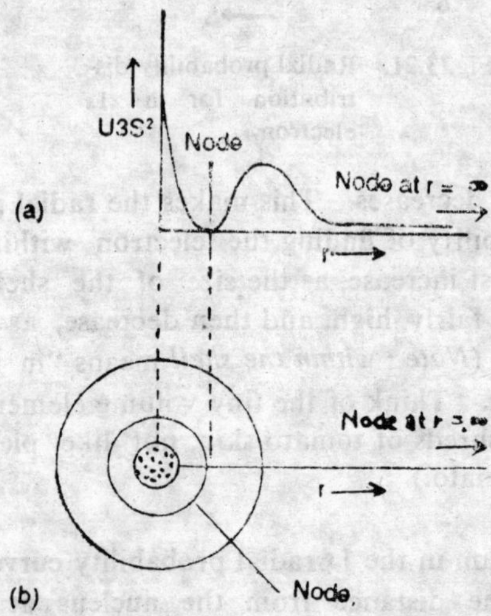

Fig. 3.22. A 2*s* orbital. (*a*) Probability-density plot (not drawn to scale). (*b*) Cross-section of electronic charge cloud.

The 1p and 3p orbitals

The shape of a boundary surface is related to the shape of the three-dimensional electron standing wave but is more properly described as either (1) the shape of the region in which the electron is most probably found, (2) the shape of the electron charge cloud, (3) the shape of the electronic probability-density distribution, or, most simply, (4) the *shape of the orbital.*

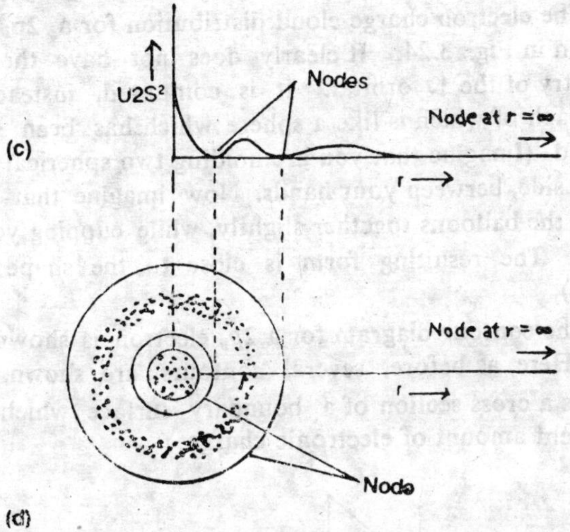

Fig. 3.23. A 3*s* orbital. (*a*) Probability-plot (not drawn to scale). (*b*) Cross section of electronic charge cloud.

The 2*p* subshell of an atom consists of *three* orbitals. Each of these is of the same energy, in the absence of external influences, at least. But each does not have the same orientasation in space. Of the three, we will first consider the $2p_y$ orbital. The wave function for an electron in this orbital can be written as

$$\psi_{2py} = Cye^{-Dr} \qquad \ldots(53)$$

Here C and D are constants, and, as before, *e* is the base for the system of natural logarithms, also a constant. Compare this equation with Eq. (3.24), namely,

$$\psi_{1s} = Ae^{-\beta r}$$

The two wave functions are similar, but the one for $2p_y$ electron has an extra y in it. This y tells us that they probability-density distribution for the $2p_y$ electron cannot be spherically symmetrical Considering usual xyz co-ordinate system (usually called three-dimensional *cartesian* co-ordinates), it means that anywhere that the value of y zero, the value of ψ, and hence ψ^2, must be zero. And where is r zero ? Anywhere on the plane defined by the x and z axes, the so-called xz plane.

The electron charge-cloud distribution for a $2p_y$ electron is shown in Fig. 3.24. It clearly does not have the spherical symmetry of the $1s$ orbital. It is composed, instead, of two *lobes*, each of which is like a sphere which has been somewhat flattened. (Imagine that you are holding two spherical balloons, side by side, berween your hands. Now imagine that you have pushed the balloons together slightly, while cupping your hands a little. The resulting form is close to the shape of a $2p$ orbital.)

The contour diagram for a $2p_y$ electron is shown in Fig. 3.25. Here, as before, several contours are shown, each of which is a cross-section of a boundary surface which encloses a different amount of electronic charge.

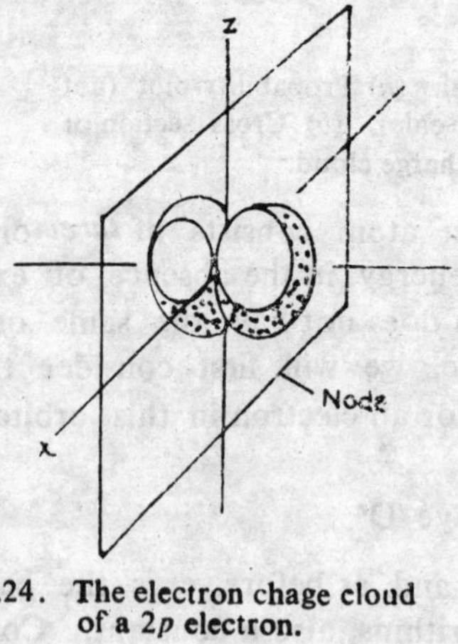

Fig. 3.24. The electron chage cloud of a $2p$ electron.

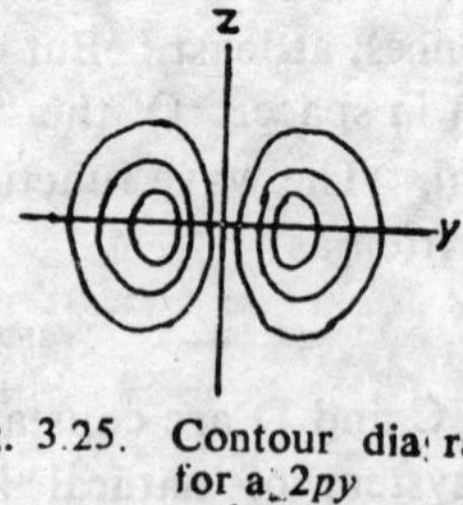

Fig. 3.25. Contour dia ram for a $2py$ orbital,

The probability of finding a $2p_y$ electron on the *xz* plane is zero. That this is true can be seen from the perspective of Fig. 3,24. This plane is a *node* which separates the two lobes

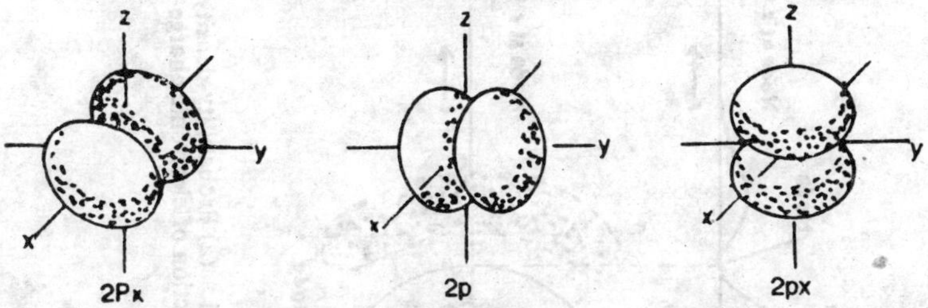

Fig. 3.26. Boundary surface diagram of the three 2p orbitals.

of the orbital. Everywhere on this nodal plane the probability of finding the electron is zero, even though there is a high-probability region on each side of the node. Thus the electron appears to spend considerable time on each side of the node but the never the node itself. This seems absurd until we remember that we are dealing with an electron which has wave-like properties. Do you recall the standing waves in a vibrating string or drum head ? These have nodes, too, and a lot of action can occur on each side of a node with none of the node itself.

The existence of nodes in probability-density distributions has prompted much conjecture about the answer to the questions. "How does the electron get across the node ?" One clever response goes, "Since the node is a geometrical plane, and since a plane has thickness, it should take zero time for the electron to pass through it !" Actually, we should not worry about the electron's crossing then ode. Just as a vibrating string has nodes and antinodes. the electronic charge distribution can have similar nodes and antinodes. Thus the electron can be thought of as being smeared out into the two lobes of a *p*-orbital with a node in between. Think of an electron as a wave.

The set of three 2p orbitals is pictured in Fig. 3.26. The boundary surfaces of these orbitals arc identical in shape, but their orientations differ. The $2p_z$ orbital has the *x* axis running through it, and the *yz* plane is a node. The *y* and *z* axes run through the $2p_y$ and $2p_z$ orbitals, respectively, and their respective nodes are the *xz* and *xy* planes.

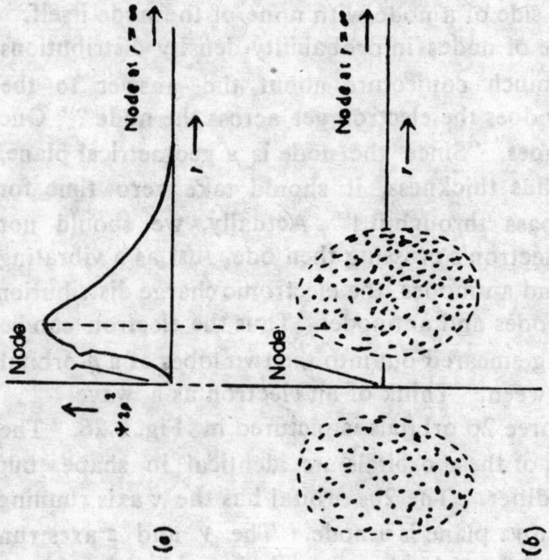

Fig. 3.27. A 2p orbital. (a) Probability-density plot. (b) Cross-section of electronic charge-cloud.

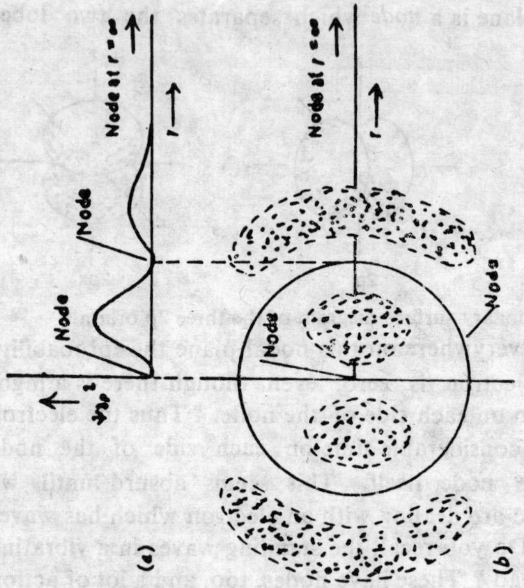

Fig. 3.28. A 3p orbital. (a) Probability-density plot. (b) Cross-section of electronic charge-cloud.

The probability-density curve for a $2p$ electron is nearly always plotted so that values of ψ^2 are shown at various distances from the nucleus along that axis which runs through the lobes. (It remains zero of course, along each of the other two axes.) Such a plot is shown in 3.27. Note that the value of ψ^2 starts at zero (at the node), increases, passes through a maximum, and finally approaches zero again as r approaches (another node). Also shown in the diagram is the cross-section of the electron charge cloud.

A $3p$ orbital is like a $2p$ orbital, except that it has an extra radial node. The charge distribution is shown in Fig. 3.27, in which both the probability density plot and the charge cloud cross-section are shown.

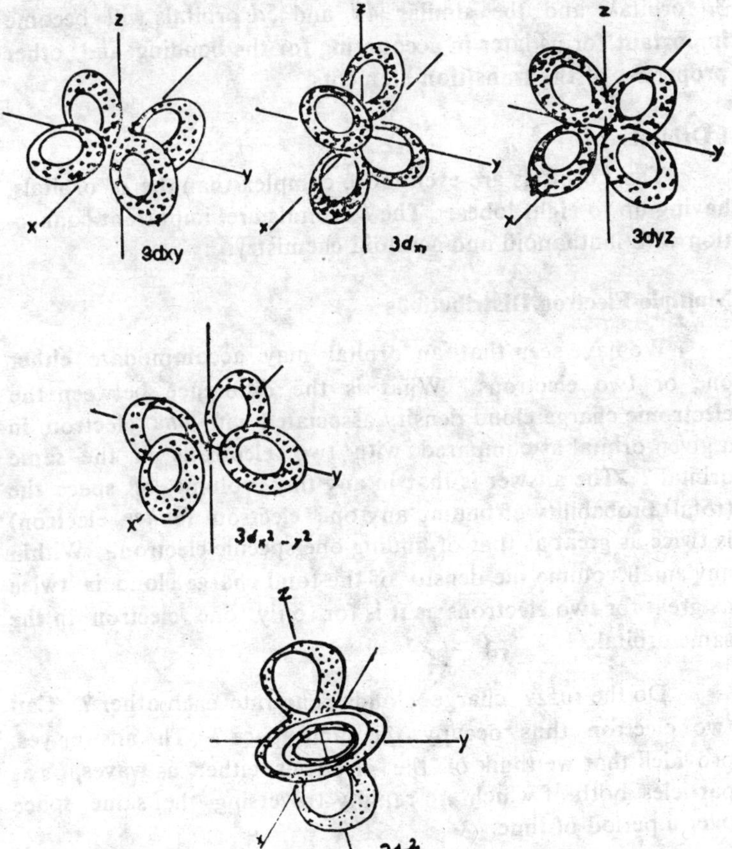

Fig. 3.29. Boundary surface diagrams of the five $3d$ orbitals.

3d Orbitals

The $3d$ subshell consists of five orbital. These are equivalent in energy (in the unperturbed atom) and are shown in charge-cloud representations in Fig. 3.29. The $3d_{xy}$ orbital consists of four lobes, each of which is located between the x and y co-ordinate axes. The $3d_{xz}$ and $3d_{yz}$ orbitals have lobes which which are similarly located between the x and z axes and between the y and z axes, respectively. The $3d_{x^2-y^2}$ orbital also has four lobes, but these are oriented so that the x and y axes poss right through the lobe centres. The $3d_{z^2}$ is different; it has a pair of lobes pointing along the z axis plus a "dough-nut" around the middle, as is shown in the diagram. These $3d$ orbitals and the similar $4d$ and $5d$ orbitals will become important for us later in accounting for the bonding and other properties of the transition elements.

f Orbitals

The f orbitals are even more complex than the d orbitals, having up to eight lobes. The f orbitals are important connection with lanthanoid and actinoid chemistry.

Multiple Electron Distributions

We have seen that an orbital may accommodate either one or two electrons. What is the difference between the electronic charge-cloud density associated with *one* electron in a given orbital as compared with two electrons in the same orbital ? The answer is that in any tiny volume of space the (total) probability of finding any one electron (*either* electron) is twice as great as that of finding one specific electron. Within any small volume the density of the total charge cloud is twiee as great for two electrons as it is for only one electron in the same orbital.

Do the fuzzy charge clouds penetrate each other ? Can two electron thus occupy the same space ? The answer yes, provided that we think of the electrons either as waves, or as particles, both of which are rapidly traversing the same space over a period of time.

What is the shape of the *total* electronic charge-cloud distribution for the $2p_z{}^12p_y{}^1$ configuration ? In this case the edges of the lobes of the $2p_z$ orbital oveslap those of the $2p_y$ orbital to form a distribution which looks like a doughnut, shown in Fig. 3.30

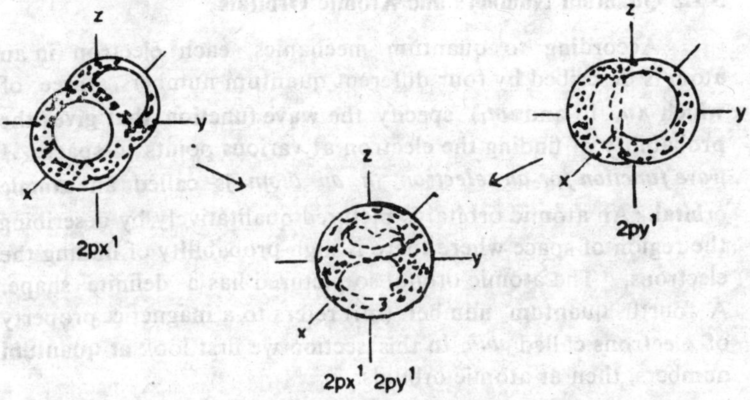

Fig. 3.30. A $2p_s12p_y1$ charge-cloud.

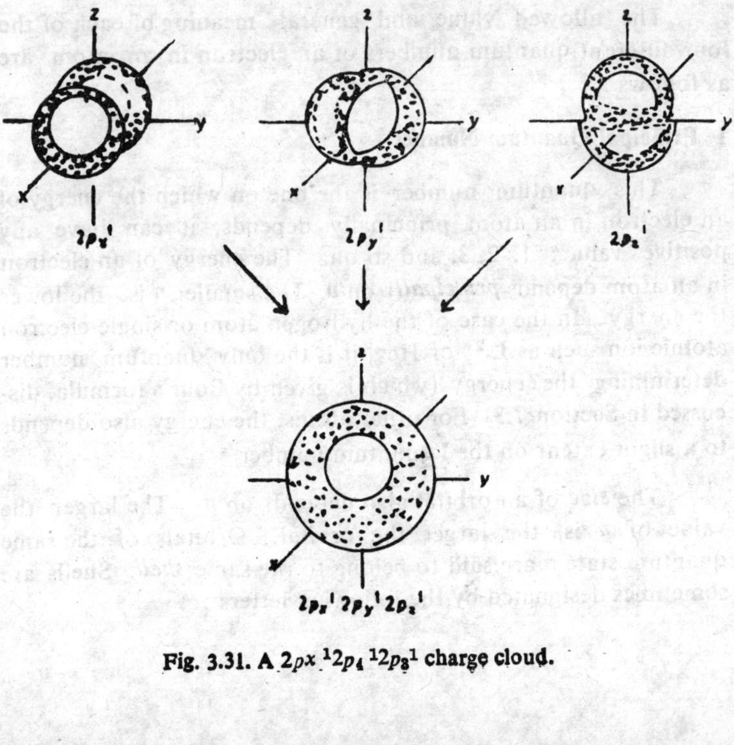

Fig. 3.31. A $2px\ ^12p_4\ ^12p_8{}^1$ charge cloud.

What is the shape of the total electronic charge-cloud distribution for the $2p_x^1 2p_y^1 2p_z^1$ configuration ? The edges of the orbitals overlap to produce an overall *spherical* charge distribution, as shown in Fig. 3.31.

3.12. Quantum Numbers and Atomic Orbitals

According to quantum mechanics, each electron in an atom is described by four different quantum numbers, three of which (n, l, and m_l) specify the wave function that gives the probability of finding the electron at various points in space. *A wave function for an electron in an atom* is called an **atomic orbital.** An atomic orbital is pictured qualitatively by describing the region of space where there is high probability of finding the electrons. The atomic orbital so pictured has a definite shape. A fourth quantum number (m_s) refers to a magnetic property of electrons called *spin*. In this section we first look at quantum numbers, then at atomic orbitals.

Quantum Numbers

The allowed value and general meaning of each of the four different quantum numbers of an electron in an atom are as follows :

1. Principal Quantum Number

This quantum number is the one on which the energy of an electron in an atom principally depends; it can have any positive value : 1, 2, 3, and so on. The energy of an electron in an atom depends *principally* on n. The smaller n is, the lower the energy. In the case of the hydrogen atom or single-electron atomic ion such as Li^{2+} or He^{+}, it is the only quantum number determining the energy (which is given by Bohr's formula, discussed in Section 7.3). For other atoms, the energy also depends to a slight extent on the l quantum number.

The *size* of an orbital also depends on n. The larger the value of n is, the larger the orbital. Orbitals of the same quantum state n are said to belong to the same *shell*. Shells are sometimes designated by the following letters :

Letter	*K*	*L*	*M*	*N...*
n	1	2	3	4...

2. Angular Quantum Number (also called Principal Quantum Number)

This quantum number distinguishes orbitals of given n having different shapes : it can have any integer value from 0 to $n-1$. Thus, within each shell of quantum number n, there are n different kinds of orbitals, each with a distinctive shape denoted by an l quantum number. For example, if an electron has a principal quantum number of 3, the possible values for l are 0, 1, and 2. Thus. within the M ($n=3$), there are three kinds of orbitals, each of which has a different shape for the region where the electron is most likely to be found. These orbital shapes will be discussed later in this section.

Table 3.1

Permissible Values of Quantum Numbers for Atomic Orbitals

n	l	m_l*	*Subshell Notation*	*Number of Orbitals in the Subshells*
1	0	0	1*s*	1
2	0	0	2*s*	1
2	1	−1,0,+1	2*p*	3
3	0	0	3*s*	1
3	1	1,0,1 1	3*p*	3
3	2	−2,−1,0,+1,+2	3*d*	5
4	0	0	4*s*	1
4	1	−1,0,+1	4*p*	3
4	2	−2,−1.0,+1,+2	4*d*	5
4	3	−3,−2,−1,0,+1,+2,+3	4*f*	7

*Any one of the *mt* quantum numbers may be associated with the *n* and *l* quantum numbers on the same line.

Orbitals of the same n but different l are said to belong to different *subshells* of a given shell. The different subshells are usually denoted by letters as follows :

Letter	*s*	*p*	*d*	*f*	*g* .....
l	0	1	2	3	4 .....

To denote a subshell within a particular shell, we write the value of the n quantum number for the shell, followed by the letter designation for the subshell. For example, $2p$ denotes a subshell with quantum numbers $n=2$ and $l=1$.

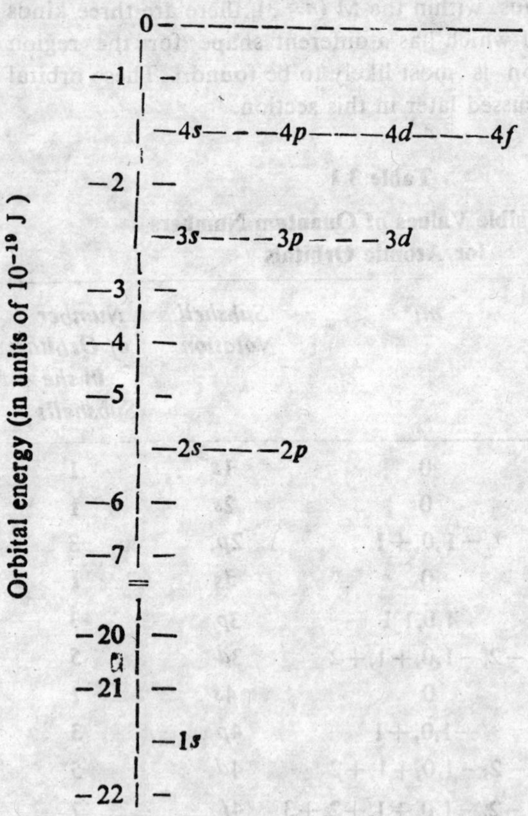

Fig. 3.32. Orbital energies of the hydrogen atom. The lines for each subshell indicate the number of different orbitals of that subshell. (Note break in the energy scale.

3. Magnetic Quantum Number

This quantum number distinguishes orbitals of given n and l—that is, of given energy and shape but having a different orientation in space; the allowed values are the integers from $-l$ to $+l$. For $l=0$ (s subshell), the allowed m_l quantum number is 0 only; there is only one orbital in the s subshell. For $l=1$ (p subshell), $m_l = -1, 0,$ and $+1$; there are three different orbitals in the p subshell. The orbitals have the same shape but different orientations in space. In addition, each orbital of a given subshell has the same energy. Note that there are $2l+1$ orbitals in each subshell of quantum number l.

4. Spin Quantum Number

This quantum number refers to the two possible orientations of the spin axis of an electron: possible values are $+\frac{1}{2}$ and $-\frac{1}{2}$. An electron acts as through it were spinning on its axis like the carth. Such an electron spin would give rise to a circulating electric charge that would generate a magnetic field. Thus, an electron behaves like a small bar magnet, with a north and south pole.

Table 1 lists the permissible quantum numbers for all orbitals through the $n=4$ shell. These values follow from the rules just given. Energies for these orbitals are shown in Fig. 3.32 for the hydrogen atom. Note that all orbitals with the same principal quantum number n have the same energy. For atoms with more than one electron, however, only orbitals in the same subshell (denoted by a given n and l) have the same energy.

Q. 1. State whether each of the following sets of quantum numbers is permissible for an electron in an atom. If a set is not permissible, explain why.

a) $n=1$, $l=1$, $m_l=0$, $m_s=+\frac{1}{2}$

(b) $n=3$, $l=1$, $m_l=-2$, $m_s=-\frac{1}{2}$

(c) $n=2$, $l=1$, $m_l=0$, $m_s++\frac{1}{2}$

(d) $n=2$, $l=0$, $m_l=0$, $m_s=1$

Solution. (a) Not permissible. The l quantum number is equal to n; it must be less than n. (b) Not permissible. The magnitude of the m_l quantum number (that is, the m_l value, ignoring its sign) must not be greater than l. (c) Permissible. (d) Not permissible. The m_s quantum number can only be $+\frac{1}{2}$ or $-\frac{1}{2}$.

4

Multielectronic Structures of Atoms

4.1. Electronic Energies

In the quantum-mechanical atom an electron is considered to behave as much like a wave as like a particle. Although it is easy to imagine waves in water, we obviously need to learn a new kind of thinking for electron waves in atoms. Furthermore, our thinking must allow us to answer three questions : (1) What are the energies of the electrons in an atom ? (2) Where is each electron ? (3) What is it doing ?

One last comment before we move forward. The electronic energy levels which we are about to describe are *hydrogenlike* energy levels; that is, they are similar to the levels obtained from quantum-mechanical calculations for a one-electron atom, hydrogen. Although this may seem like an oversimplification, in practice is works out very well.

Shells

In order to describe electronic energies in an atom, it is convenient to organize them into sets each of which is called a *shell*. A shell is therefore a collection of one or more quantized energy levels.

Shells are named in two different ways : (1) by using the letters K, L, M, N, O, P.,..., and (2) by specifying the number n of the shell, where n can be 1, 2, 3, 4, 5, 6,.. Electrons in the K shell ($n=1$) have the lowest permitted energy.

The maximum electron population of a shell can be shown both theoretically and experimentally to be equal to $2n^2$. Thus the K shell ($n=1$) can hold up to two electrons, the L shell ($n=2$) up to eight electrons, and so on. The shells which consist of the lower electronic energies in an atom are schematically shown at the left of Fig. 4.1.

Subshell

Not all electrons in a given shell have the same energy. Each shell is thus considered to consist of, or contain, one or more *subshells.* (The number of subshells in any given shell is equal to n). Within a given shell the subshells are named in order of *increasing energy* : *s, p, d, f, g, h, i*,... (Only the first four, *s, p, d,* and *f*, are needed for describing atoms in their

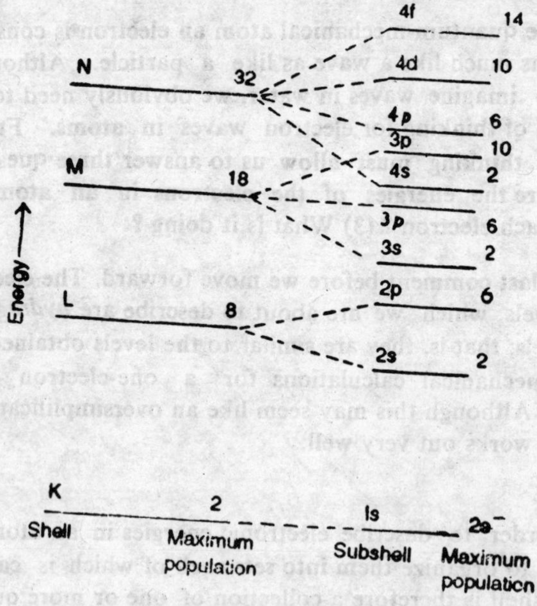

Fig. 4.1 Shells and subshells.

ground states). A specific subshell is named by using the number which designates the shell, plus the letter which indicates the subshell, such as, for example, $2p$, $4d$, $5f$, etc.

As stated above the number of subshells into which a given shell is subdivided depends on the quantum number n. The K shell ($n=1$) has but *one* subshell, called the $1s$ subshell. The L shell ($n=2$) has *two* : $2s$ and $2p$; the M shell ($n=3$) *three* : $3s$, $3p$, and $3d$; the N shell ($n=4$) *four* : $4s$, $4p$, $4d$, and $4f$, etc.

The *maximum* electronic population of an individual subshell depends upon whether the subshell is *s*, *p*, *d*, or *f*. There can be no more than 2 electrons in any *s* subshell, 6 in any *p* subshell, 10 in any *d* subshell, and 14 in any *f* subshell.

Confirmation of the existence of subshells comes from studies of atomic spectra. Subdivision of shells into subshells occurs because electrons in an atom are not independent of each other but interact electrically. Even hydrogen atoms with their single electrons show this splitting of energy levels in the presence of an electric field. (This is called the *Stark effect*.)

Energy relationships among the subshells of the first four shells (K, L, M, and N) are shown at the right of Fig. 4.1. Note the energy overlap of the $4s$ subshell with the $3d$. The $4s$ subshell actually is of *lower* energy than the $3d$. This is partially a result of the bunching up of levels at higher energies (toward the top of the diagram).

Orbitals

In an isolated atom all electrons in a given subshell have the same energy. Under perturbing influences, however, their energies may be further split into sublevels called *orbitals*. Such splitting occurs, for example, in the presence of a magnetic field. (This is called the *Zeeman effect*).

The word *orbital*, is, like *shell*, somewhat unfortunate. It is derived from the word *orbit*, as used by Bohr, but it does not

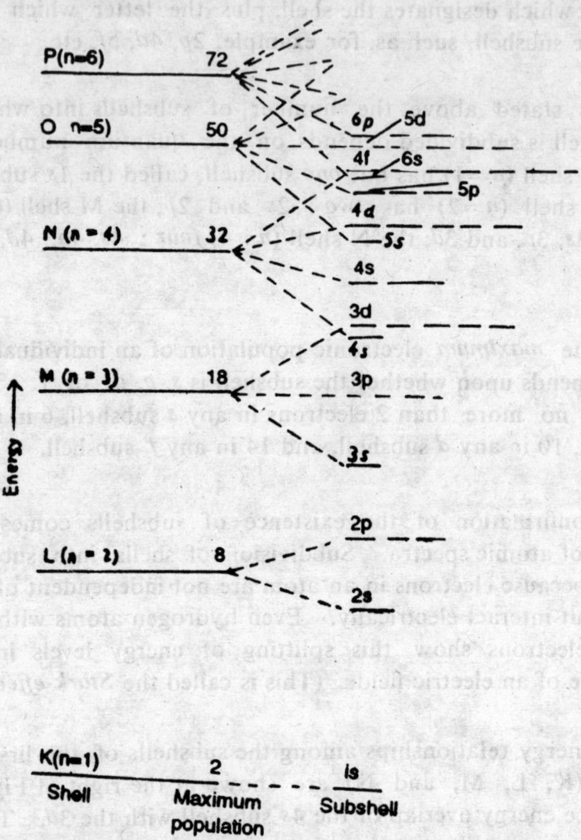

Fig. 4.2. Filling diagram (partial).

in any way signify an orbit. For now think of an orbital simply as an energy level. Later (Sec. 8-4) we will see that *orbital* can be taken to mean a region of space in which an electron of a certain energy is likely to be found.

The number of orbitals into which a subshell is subdivided or split depends on its identity. The number of orbitals in an *s*, *p*, *d*, or *f* subshell is 1, 3, 5, or 7, respectively. Comparing these numbers with the maximum population of each kind ol subshell (2, 6, 10, or 14, respectively), we see that *each orbital has a maximum population of two electrons.*

For the lower energy levels the subdivision of subshells into orbitals is shown schematically in Fig. 4.2. In this diagram the orbitals of a given subshell are all shown at the same energy, as they would exist in an isolated atom not subject to exernal fields.

Etctron Spin

Theoretical calculations and spectroscopic data show that each orbital may contain one or two electrons. We may also refer to an *empty*, or *unoccupied*, orbital, which is a potentially available energy level with no electrons in it. Furthermore, both theory and experiment support the idea that when two electrons occupy the same orbital, they *have different spins.* What does this mean ? An interesting experiment performed in 1921 by Otto Stern and Walter Gerlach showed that an electron acts as if it could *spin* in either of two opposite directions. The Stern-Gerlach experiment, shown in Fig. 4.3, consists of passing a beam of metal atoms (Stern and Gerlach used silver, sodium, potassium, and others) through an inhomogeneous (nonuniform) magnetic field. With these metals the beam divides into two components, one being deflected up and the other down. The principles of physics tell us that a particle passing through such a magnetic field will be deflected (1) if the particle is *charged* and (2) if it is *spinning.* If a spinning electron were passed through such a magnetic field, wewould expect it to be deflected either up or down, depending on the direction of its spin. But what about a sodium atom ? A sodium atom has 11 electrons. Five of these spin in one direction, and six in the other. The five which spin in the first direction have their magnetic effects exactly canceled by five of those which spin in the opposite direction. Thus the magnetic effects of ten electrons are self-canceling. But the last electron is uncompensated for and causes the entire atom to be deflected. Since half of the sodium atoms have their "eleventh" electrons spinning one way, and half the other, two beams of atoms emerge from the magnetic field. There are other experimental methods of showing that the electron has spin, but the Stern-Gerlaeh experiment is the most direct.

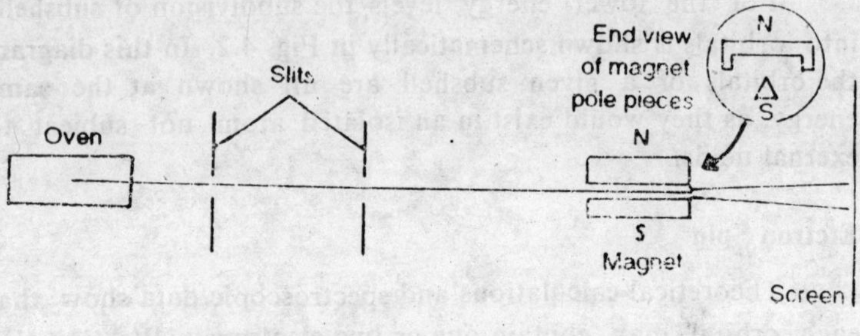

Fig. 4.3. Stern-Gerlach experiment.

Each orbital in an atom can accommodate a maximum of two electrons, one spinning one way and the other spinning in the opposite way. We say that two such electrons are paired and have opposing, opposite, or antiparallel spins. The spins of electrons contribute to the magnetic behavior of matter. This is because any spinning or rotating object which is electrically charged generates a magnetic field, that is, acts as if it were a magnet. Substances which contain atoms with orbitals which are half-filled (contain only one electron) are weakly attracted into a magnetic field, that is "toward a magnet". Such a weak attraction is called paramagnetism and is evidence of the existence of unpaired electrons in a substance.

4.2. Pauli Exclusion Principle

Not all of the conceivable arrangements of electrons among the orbitals of an atom are physically possible. The **Pauli exclusion principle**, which summarizes experimental observations, is a rule stating that no two electrons in an atom can have the same four quantum numbers. If one electron in an atom has the quantum numbers $n=1$, $l=0$, $ml=0$, ond $m_s=+\frac{1}{2}$ no other electron can have the same four quantum numbers. In other words, you cannot place two electrons with the same value of ml in a $1s$ orbital. The orbital diagram

↑ ↑
s

does not represent a possible arrangement of electrons.

Because there are only two possible values of m_s an orbital can hold no more than two electrons and than only if two electrons have different spins.

In an orbital diagram, an orbital with two electrons has to be written with arrows pointing in opposite directions. The two electrons are said to have opposite spins.

You can see that each subshell holds a maximum of twice as many electrons as the number of orbitals in the subshell. Thus, a $2p$ subshell, which has three orbitals (with $m_l = -1$, 0, and $+1$), can hold a maximum of six electrons. The maximum number of electrons in various subshells is given in the following table.

Subshell	*Number of Orbitals*	*Maximum Number of Electrons*
s ($l = 0$)	1	2
p ($l = 1$)	3	6
d ($l = 1$)	5	10
f ($l = 3$)	7	14

Example 4.1. Which of the following orbital diagrams or electron configurations are possible and which are impossible, according to the Pauli exclusion principle ? Explain.

(a) ↓ ↑   ↑ ↓   ↑
$1s$ $2s$ $2p$

(b) ↑ ↓
$1s$ $2s$ $2p$

(c) ↑ ↓   ↑   ↑ ↑
$1s$ $2s$ $2p$

(d) $1s^1 2s^1$

(e) $1s^2 2s^2 2p^7$

(f) $1s^2 2s^2 2p^6 3s^2 3p_6 3d^8 4s^2$

Solution

(a) Possible orbital diagram. (b) Impossible orbital diagram; there are three electrons in the $2s$ orbital. (c) Impossible obrital diagram; there are two electrons in a $2p$ orbital with the same spin. (d) Impossible electron configuration; there are three electrons in the $1s$ subshell (one orbital). (e) Impossible electron configuration; there are seven electrons in the $2p$ subshell (which can hold only six electrons). (f) possible Note that the $3d$ subshell can hold as many as ten electrons.

4.3 Building-up Principle and the Periodic Table

Every atom has an infinite number of possible electron configurations. The configuration associated with the lowest energy level of the atom corresponds to a quantum-mechanical state called the ground state. Other configurations correspond to excited states, associated with energy levels other than the lowest. For example, the ground state of the sodium atom is known from experiment to have the electron configuration $s^2 2s^2 2p^6$ $3s^1$. The electron configuration $1s^2 2s^2 2p^6 3p^1$ represents an excited state of the sodium atom.

The chemical properties of an atom are related primarily to the electron configuration of its ground state. Table 4.1 lists the experimentally determined ground-state electron configurations of atom Z=1 to Z=36.

Building-up Principle (Aufbau Principle)

Most of configuration in Table 4.1 can be explained in terms of the building-up principle (or Aufbau principle), a scheme use to reproduce the electron configurations of the ground states of atom by successively filling subshells with electrons in a specific order (the building-up order). Following this principle, you obtain the electron configuration of an atom by successive.y filling subshells in the following order (the building-up order) ; $1s$, $2s$, 2p, $3s$, $3p$, $4s$, $4d$, $4p$, $5s$, $4d$, $5p$, $6s$, $4f$, $4d$, $6p$, $7s$, $5f$. This order produces the experimentally determined electron configurations (with some exceptions, which we will discuss later). You need not memorize this order.

As you will see, you can very easily obtain it from the periodic table. Before we describe how to do that, we need to discuss this building-up order and see how the electron configurations very as we progress through the periodic table.

The building-up order corresponds for the most part to increasing energy of the subshells. You might expect this. By filling orbitals of lowest energy first, you usually get the lowest total energy (ground state) of the atom. Recall that the energy of an orbital depend only on the quantum numbers n and l. (The energy of the H atom, however, depends only on n), Orbitals with the same n and l but different m_1 – that is, different orbitals of the same subshell – have the same energy. The energy depends primarily on n, increasing with its value. Thus, a 3*s* orbital has greater energy than a 2*s* orbital because the value of n is greater. Except for the H atom, the energies of orbitals with the same n increase with the l quantum number. A 3*p* orbital has slightly greater energy than a 3*s* orbital because l is greater. The orbital of lowest energy is 1*s*: the next higher once are 2*s* and 2*p*, then 3*s* and 3*p*. The order of these subshells by energy, from 1*s* to 3*p*. follows the building-up order as we listed it earlier.

We subshells have nearly the same energy, however, the building-up order is not strictly determined by the order of their energies. The ground-state configurations, which we are trying to predict by the building-up order, are determined by the total energies of the atoms. The total energy of an atom depends not only on the energies of the subshells but also on the energies of interaction among the different subshells of an atom. Thus, for all elements with $Z = 21$ or greater, the energy of the 3*d* subshells is lower than the energy of the 4*s* subshells (Figure 4.7). which is opposite to the building-up order. It is the building-up order that you need in order to predict the electron configurations of the ground states of atom.

Now you can see how to produce the electron configurations of Table 4.1 using the building-up principle, Remember that the number of electrons in a neutral atom equals the atomic number Z (the nuclear charge is +Z). In the case of the simplest atom, hydrogen (Z = 1), you obtain the ground state by placing the single electron into the 1*s* orbital, giving the configuration 1*s*¹ (this is read as "one-essone"). Now you

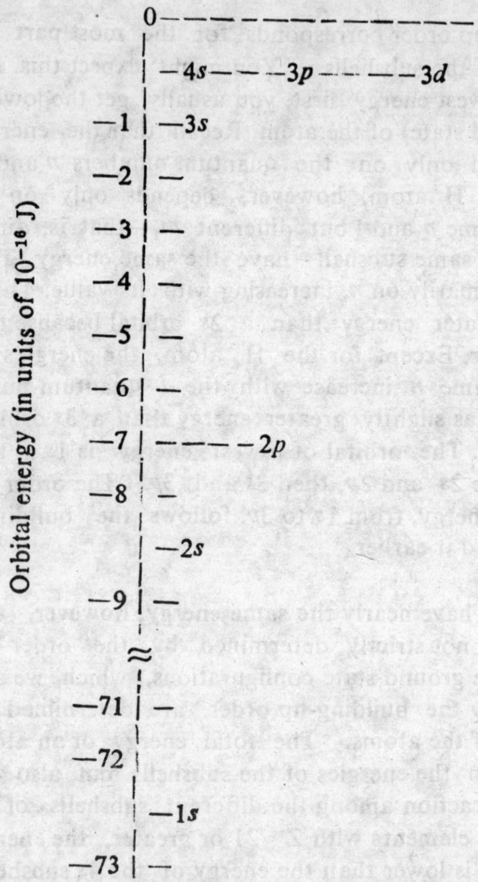

Fig. 4.4. Orbital energies for the scandium atom (Z−21). Note that in the scandium atom, unlike the hydrogen atom, the subshells for each *n* are spread apart in energy. Thus, the 2p energy is above the 2s. Similarly, the *n*=3 subshells are spread to give the order $3s < 3p < 3d$. The 3d subshell energy is now just below the 4s. (V4lues for this figure were calculated from theory by Charlotte F. Fischer, Vanderbilt University).

go to hilium (Z=2). The first electron goes into the 1*s* orbital as in hydrogen, followed by the second electron, because any orbital can hold two electrons. The configurations is $1s^2$. Filling the $n=1$ shell creates a very stable configuration and, as a result, gives rise to a chemically unreactive atom.

You continue this way through the elements, each time increasing Z by one and adding another electron. You obtain the configuration of an atom from that of the preceding element by adding an electron into the next available orbital, following the building-up order. In lithium (Z=3), the first two electrons gives the configuration $1s^2$, like helium, but the third electron goes into the next higher orbital in the building-up order, because the 1*s* orbital is now filled. This gives the configuration $1s^2\,2s^2$. In beryllium (Z=4), the fourth electron fiills the 2*s* orbital, giving the configuration $1s^22s^2$.

If you use the abbreviation [He] for $1s^2$, the configuration are

Z=3	lithium	$1s^22s^1$	or	[He]$2s^2$
Z=4	beryllium	$1s^22s^2$	or	[He] $2s^2$

With boron (Z=5), the electrons begin fiilling the 2 *p* subshell. You get

Z=5	boron	$1s^22s^22p^1$	or	[He] $2s^22p^2$
Z=6	carbon	$1s^22s^22p^2$	or	[He]es^22p^2
⋮				
Z=10	neon	$1s^22s2p^2$	or	[He] $2s^2\,2p^6$

Having filled the 2*p* subshell, you again find a particularly stable configuration. Neon is chemically unreactive as a result.

With sodium (Z=11), the 3*s* orbital begins to fill. Using the abbreviation [Ne] for $1s^22s^22p^6$, you have

Z=11	sodium	$1s^22s^22p^63s^1$	or	[Ne]$3s^1$
Z=12	magnesium	$1s^22s^22p^63s^2$	or	[Ne] $3s^2$

Then the 3*p* subshell begins to fill.

Z=13	aluminium	$1s^2\,2s^22p^63s^23p^1$	or	[Ne] $3s^23p^2$
⋮				
Z=18	argon	$1s^22s^22p^63s^23p^6$	or	[Ne] $3s^23p^6$

With the $3p$ subshell filled, a stable configuration has been attained ; argon is an unreactive element.

Now the $4s$ orbital begins to fill. You get [Ar] $4s^1$ for potassium (Z=19) and [Ar] $4s^2$ for calcium [Ar]=$1s^22s^22p^6$ $3s^23p^6$). At this point the $3d$ subshell begins to fill. You get [Ar] $3d^14s^2$ for scandium (Z=21), [Ar] $3d^2\ 4s^2$ for titanium (Z=22),and [Ar] $3d^3\ 4s^2$ for vanadium (Z=23). Note that we have written the configurations with subshells arranged in order by shells. This generally places the subshells in order by energy and puts the subshells that are involved in chemical reactions at the far right.

Let us skip to zinc (Z=30). The $3d$ subshell has filled ; configuration is [Ar] $3d^{10}\ 4s^2$. Now the $4p$ subshell begins to fill, starting with gallium (Z=31), configuration [Ar] $3d^{10}\ 4s^2$ $4p^2$, and ending with krypton (Z=36). configuration [Ar] $3d^{10}$ $3s^2\ 4p^6$.

Electron Canfigurations and the Periodic Table

By this time you can see a pattern develop among the ground-state electron configurations of the atoms. You can see that this pattero explains the periodic table, which was briefly described in Section 2.5. Consider helium, neon, argon and krypton, which are elements in Group VIII A of the periodic table Neon, argon, and krypton have configurations in which a p subshell has just filled (helium has filled $1s$ subshell, no $1p$ subshell is possible).

helium	$1s^2$
neon	$1s^2\ 2s^2\ 2p^2$
argon	$1s^2\ 2s^22p^6\ 3s^23p^6$
krypton	$1s^22s^22p^63s^23^{86}3d^{10}4s^24p^6$

These elements are the first members of the group of elements called *noble gases* because of their relative unreactivity.

Look now at the configurations of beryllium, magnesium, and calcium, members of the group of alkaline earth metals (Group IIA), which are similar. moderately reactive elements.

Table 4.1

Eelectronic Configurations of the Elements

Element	Z		Shell (n in parentheses) K (1)	L (2)	M (3)	N (4)	O (5)	P (6)	Q (7)
			s	s p	s p d	s p d f	s p d f	s p d f	s
H	1		1						
He	2		2						
Li	3		2	1					
Be	4		2	2					
B	5		2	2 1					
C	6		2	2 2					
		(He)							
N	7		2	2 3					
O	8		2	2 4					
F	9		2	2 5					
Ne	10		2	2 6					
Na	11		2	2 6	1				
Mg	12		2	2 6	2				

(Contd.)

Al	13		2	2	6	2	1			
Si	14	(Ne)	2	2	6	2	2			
P	15		2	2	6	2	3			
S	16		2	2	6	2	4			
Cl	17		2	2	6	2	5			
Ar	18		2	2	6	2	6			
K	19		2	2	6	2	6		1	
Ca	20		2	2	6	2	6		2	
Sc	21		2	2	6	2	6	1	2	
Ti	22		2	2	6	2	6	2	2	
V	23		2	[illegible]	6	2	6	3	2	
Cr	24		2	2	6	2	6	5	1	
Mn	25		2	2	6	2	6	5	2	
Fe	26		2	2	6	2	6	6	2	
Co	27		2	2	6	2	6	7	2	
Ni	28	(Ar)	2	2	6	2	6	8	2	
Cu	29		2	2	6	2	6	10	1	
Zn	30		2	2	6	2	6	10	2	
Ga	31		2	2	6	2	6	10	2	
Ge	32		2	2	6	2	6	10	2	2
As	33		2	2	6	2	6	10	2	3
Se	34		2	7	6	2	6	10	3	4
Br	35		2	2	6	2	6	10	2	5
Kr	36		2	2	6	2	6	10	2	6

Element	Z	K	L	M	N	O	P
Rb	37	2	2 6	2 6 10	2 6	1	
Sr	38	2	2 6	2 6 10	2 6	2	
Y	39	2	2 6	2 6 10	2 6 1	2	
Zr	40	2	2 6	2 6 10	2 6 2	2	
Nb	41	2	2 6	2 6 10	2 6 4	2	
Mo	42	2	2 6	2 6 10	2 6 5	1	
Tc	43	2	2 6	2 6 10	2 6 6	1	
Ru	44	2	2 6	2 6 10	2 6 7	2	
Rh	45	2	2 6	2 6 10	2 6 8	1	
	(Kr)						
Pd	46	2	2 6	2 6 10	2 6 10		
Ag	47	2	2 6	2 6 10	2 6 10	1	
Cd	48	2	2 6	2 6 19	2 6 10	2	
In	49	2	2 6	2 6 10	2 6 10	2 1	
Sn	50	2	2 6	2 6 10	2 6 10	2 2	
Sb	51	2	2 6	2 6 10	2 6 10	2 3	
Te	52	2	2 6	2 6 10	2 6 10	2 4	
I	53	2	2 6	2 6 10	2 6 10	2 5	
Xe	54	2	2 6	2 6 10	2 6 14	2 6	
Cs	55	2	2 6	2 6 10	2 6 10	2 6	1
Ba	56	2	2 6	2 6 10	2 6 10	2 6	2
La	57	2	2 6	2 6 10	2 6 10	2 6 1	2
Ce	58	2	2 6	2 6 10	2 6 10 1	2 6 1	2

Pr	59		2	2	6	2	6	10	2	6	10	2	2	6		2	
Nd	60		2	2	6	2	6	14	2	6	10	3	2	6		2	
Pm	61		2	2	6	2	6	10	2	6	10	4	2	6		2	
Sm	62		2	2	6	2	6	10	2	6	10	5	2	6		2	
Eu	63		2	2	6	2	6	10	2	6	10	6	2	6		2	
Gd	64		2	2	6	2	6	10	2	6	10	7	2	6	1	2	
Tb	65		2	2	6	2	6	10	2	6	10	7	2	6		2	
Dy	66		2	2	6	2	6	10	2	6	10	9	2	6		2	
Ho	67		2	2	6	2	6	10	2	6	10	10	2	6		2	
Er	68		2	2	6	2	6	10	2	6	10	11	2	6		2	
Tm	69		2	2	6	2	6	10	2	6	10	12	2	6		2	
Yb	75		2	2	6	2	6	10	2	6	10	14	2	6		2	
Lu	71		2	2	6	2	6	10	2	6	10	14	2	6	1	2	
Hf	72		2	2	6	2	6	10	2	6	10	14	2	6	2	2	
		(Xe)															
Ta	73		2	2	6	2	6	10	2	6	10	14	2	6	3	2	
W	74		2	2	6	2	6	10	2	6	10	14	2	6	4	2	
Re	70		2	2	6	2	6	10	2	9	10	14	2	6	5	2	
Os	76		2	2	6	2	6	10	2	6	10	14	2	6	6	2	
Ir	77		2	2	6	2	6	10	2	6	10	14	2	6	7	2	
Pt	78		2	2	6	2	6	10	2	6	10	14	2	6	8	1	
Au	79			2	6	2	6	10	2	6	10	14	2	6	10	1	
Hg	80		2	2	6	2	6	10	2	6	10	14	2	6	10	2	
Ti	81		2	2	6	2	6	10	2	6	10	14	2	6	10	2	1
Pb	82		2	2	6	2	6	10	2	6	10	14	2	6	10	2	2

Bi	83	2	2 6	2 6 10	2 1 10 14	2 6 10	2 3	
Po	84	2	2 6	2 6 10	2 6 10	2 6 10	2 4	
At	85	2	2 6	2 6 10	2 6 10	2 6 10	2 5	
Rn	86	2	2 6	2 6 10	2 6 19	2 6 10	2 6	
Fr	87	2	2 6	2 6 10	2 6 10 14	2 6 10	2 6	1
Ra	88	2	2 6	2 6 10	2 6 10 14	2 6 10	2 6	2
Ac	89	2	2 6	2 6 10	2 6 10 14	2 6 10	2 6 1	2
Th	90	2	2 6	2 6 10	2 6 10 14	2 6 10	2 6 2	2
Pa	91	2	2 6	2 6 10	2 6 10 14	2 6 10 2	2 6 1	2
U	92	2	2 6	2 6 10	2 6 10 14	2 6 10	2 6 2	2
Np	93	2	2 6	2 6 10	2 6 10 14	2 6 10 4	2 6 1	2
Am	94	2	2 6	2 6 10	2 6 10 14	2 5 10 6	2 6	2
Cm	95	2	2 6	2 6 10	2 6 10 14	2 6 10 7	2 6	2
Bk	96	2	2 6	2 6 10	2 6 10 14	2 6 10 8	2 6	2
Cf	97	2	2 6	2 6 10	2 6 10 14	2 6 10 9	2 6	2
Es	98	2	2 6	2 6 10	2 6 10 14	2 6 10 10	2 6	2
Fm	99	2	2 6	2 6 10	2 6 10 14	2 6 10 11	2 6	2
Md	100	2	2 6	2 6 10	2 6 10 14	2 6 10 12	2 6	1
No	101	2	2 6	2 6 10	2 6 10 14	2 6 10 13	2 6	2
Lr	102	2	2 6	2 6 10	2 6 10 14	2 6 10 14	2 6	2
Rf	103	2	2 6	2 6 10	2 6 10 14	2 6 10 14	2 6 2	2
Ha	105	2	2 6	2 6 10	2 6 10 14	2 6 10 14	2 6 3	2
	106	2	2 6	2 6 10	2 6 10 14	2 6 10 14	2 6 3	2

beryllium	$1s^22s^2$	or [He]$2s^2$
magnesium	$1s^22s^2nh^63s^2$	or [He]$3s^2$
calcium	$1s^22s^22p^63s^23p^64s^2$	or [Ar]$4s^2$

Each of these configurations consist of a **noble gas care**, that is an inner-shell configuration corresponding to one of the noble gases, plus to outer electrons with an ns^2 configuration.

The elements boron, aluminum, and gallium (Group IIIA) also have similarities. Their configurations are

boron	$1s^22s^22p^1$	or [He]$2s^22p^1$
aluminum	$1s^22s^22p^63s^23p^1$	or [Ne]$3s^23p^1$
gallium	$1s^22s^22p^63s^23p^63d^{10}4s^24p^1$	or [Ar]$3d^{10}4s^24p^1$

Boron and aluminum have noble-gas cores plus three electrons with the configuration ns^2np^1. Gallium has an filled $3d$ subshell. The noble-gas core together with $(n-l)a^{10}$ electrons is often referred to as a *pseudo-noble-gas core*, because these electrons usually are not involved in chemical reactions.

An electron in an atom outside the noble-gas or pseudo-noble-gas core is called a *valence electron*. It is such electrons that are primarily involved in a chemical reactions, and similarities among the configurations of valence electrons (the valence-shell configurations) account for similarities of the chemical properties among groups of elements.

Exceptions to the Building-Up Principle

As we have said, the building-up principle reproduces most of the ground-state configurations correctly. There are some exceptions, however, and chromium ($Z=24$) is the first one we encounter. The building-up principle predicts the configuration [Ar]$3d^44s^2$, though the correct one is found experimentally to be [Ar]$3d^54s^1$. These two configurations are actually very close in total energy because of the closeness in energies of the $3d$ and $4s$ orbitals. For that reason, small effects can influence which of the two configurations is actually lower in energy. Copper ($Z=29$) is another exception to the building-up principle, which predicts the configuration

[Ar]$3d^9 4s^2$, although experiment shows the ground state to be the configuration [Ar]$3d^{10} 4s^1$.

We need not dwell on these exceptions beyond noting that they occur. The point to remember is that the configuration predicted by the building-up principle is very close in energy to the ground-state configuration (if it is not the ground state). Most of the qualitative conclusions regarding the chemistry of an element are not materially affected by arguing from the configuration given by the building-up material.

4.4. Orbital Diagrams of Atoms; Hund's Rule

In discussing the ground state of atoms, we have not yet described how the the electrons are arranged within each subshell. They may be several different ways of arranging electrons in a particular configuration. Consider the carbon atom ($Z=6$) with the ground-state configuration $1s^2 2s^2 2p^2$. Three possible arrangements are given in the following orbital diagrams.

Diagram 1 : ↑↓ ↑↓ ↑ ↑
1*s* 2*s* 2*p*

Diagram 2 : ↑↓ ↑↓ ↑↓
1*s* 2*s* 2*p*

Diagram 3 : ↑↓ ↑↓ ↑↓
1*s* 2*s* 2*p*

These orbital diagrams show different state of the carbon atom. Each state has a different energy and, as you will see, different magnetic characteristics.

Hund's Rule

In about 1927, Friedrich Hund discovered an emprical rule determining the lowest-energy arrangement of electrons in a subshell. *Hund's rule* states that the lowest-energy arrangement of electrons in a subshell is obtained by putting electrons into separate orbitals of the subshell with the same spin before pairing electrons. Let us see how this would apply to the case of the carbon atom, whose ground-state configuration is $1s^2 2s^2 2p^2$. The first four electrons go into the 1*s* and 2*s* orbitals.

↑↓ ↑↓
1*s* 2*s* 2*p*

The next two electrons go into separate 2*p* orbitals, with both electrons having the same spin, following Hund's rule,

↑↓ ↑↓ ↑↑
1*s* 2*s* 2*p*

Thus, the orbital diagram corresponding to the lowest energy is the one we previously labelled as Diagram 1.

To apply Hund's rule to the oxygen atom, whose ground-state configuration is $1s^27s^22p^4$, we place the first seven electrons as follows :

↓↑ ↓↑ ↑↑↑
1*s* 2*s* 2*p*

The last electron is paired with one of the 2*p* electrons to give a doubly occupied orbital. The orbital diagram for the ground state of the oxygen atom is

↑↓ ↑↓ ↑↓↑↑
1*s* 2*s* 2*p*

In the next example, Hund's rule is used to determine the orbital diagram for the ground state of a more complicated atom. Orbital diagrams for the ground states of the first ten elements are shown in Table 2.

Example 4.2. Write an orbital diagram for the ground state of the from atom.

Solution. From the building-up principle, you determine that the electron configuration of the iron atom is $1s^22s^22p^63s^2$-$3p^63d^64s^2$. All the subshells except the 3*d* are filled. In placing the six electrons in the 3*d* subshell, you note that the first five go into separte 3*d* orbitals with their spin arrows in the same direction. The sixth electron must doubly occupy a 3*d* orbital. The orbital diagram is

↑↓ ↑↓ ↑↓↑↓↑↓ ↑↓ ↑↑↑↓↑↑ ↑↓↑↑↑↑ ↑↓
1*s* 2*s* 2*p* 3*s* 3*p* 3*d* 4*s*

You can write this diagram in abbreviated from using [Ar] for the argonlike core of the iron atom :

[Ar] ↑↓ ↑ ↑ ↑ ↑↓
3*d* 4*s*

Magnetic Properties of Atoms

The magnetic properties of a snbstance can reveal certain information about the arrangement of electrons in atom (or molecule). Although an electron in an atoms behaves like a small magnet, the magnetic attractions from two electrons that are opposite in spin cancel each other. As a result, an atom that has only doubly occupied orbitals has no net spin magnetism. However, an atom in which there are *unpaired* electron—that is, in which there is an excess of one kind of spin - does exhibit a net magnetism.

The magnetic properties of an atom can be observed. The most direct way is to determine whether the atomic substance is attracted to the field of a strong magnet. A *paramagnetic substance* is a substance that is weakly attracted by a magnetic field, and this attraction is generally the result of unpaired electrons. For example, sodium vapour has been found experi-

Table 4.2
Orbital Diagrams for the Ground States Atoms from Z=1 to Z=10

Atom	*Z*	*Configuration*	*Orbital Diagrams* 1*s*	2*s*	3*p*
Hydrogen	1	$1s^1$			
Helium	2	$1s^2$			
Lithium	3	$1s^22s^1$			
Beryllium	4	$1s^22s^2$			
Boron	5	$1s^22s^22p^1$			
Carbon	6	$1s^22s^22p^2$			
Nitrogen	7	$1s^22s^22p^3$			
Oxygen	8	$1s^22s^22p^4$			
Fluorine	9	$1s^22s^22p^5$			
Neon	10	$1s^22s^22p^6$			

mentally to be paramagnetic. The explanation is that the vapor consists primarily of sodium atoms, each containing an unpaired electron (the configuration is [Ne]$3s^1$). A *diamagnetic substance* is a substance that is not attracted by a magnetic field or is very slightly repelled by such a field, This property generally means that the substance has only paired electrons. Mercury vapour is found experimentally to be diamagnetic. The explanation is that mercury vapour consists of mercury atoms, each with the electron configuration Xe—$4f^{14}5d^{10}6s^2$, which has only paired.

We expect the orbital diagrams presented at the beginning of this section for $1s^22s^22p^2$ configuration of the carbon atom to have different magnetic properties. Diagram 1, which is predicted by Hund's rule to be the ground state, would give a magnetic atom, whereas the other diagrams would not. If we could prepare a vapour of carbon atom, it could be attracted to a magnet (it should be paramagnetic). It is difficult to prepare a vapour of free carbon atoms in sufficient concentration to observe a result. However, the spectrum of carbon atoms can be obtained easily from dilute vapour. From an analysis of this spectrum, it is possible to show that the ground-state atom is magnetic, which is consistent with the prediction of Hund's rule.